Value Proposition

Jian Yuchi

Value Proposition

A New "Long March" & E^3 Economy on China's Integrated Development of Internet, Big Data, AI and Manufacturing Industry

北京交通大学出版社
http://press.bjtu.edu.cn

Springer

Jian Yuchi
Ministry of Science
and Technology of China
High and New Technology
Development and Industrialization
Beijing, China

Translated by
Jian Zuo
School of Architecture
and Built Environment
University of Adelaide
Adelaide, Australia

ISBN 978-981-15-5024-9 ISBN 978-981-15-5025-6 (eBook)
https://doi.org/10.1007/978-981-15-5025-6

Jointly published with Beijing Jiaotong University Press
The print edition is not for sale in China Mainland. Customers from China Mainland please order the print book from: Beijing Jiaotong University Press.

This Springer imprint is published by the registered company Springer Nature Singapore Pte Ltd.
The registered company address is: 152 Beach Road, #21-01/04 Gateway East, Singapore 189721, Singapore

Foreword by Zhenhua Zhao

It is not necessary to have met with each other before we become the best friends. Indeed, the communication of spirit and ideas play the most critical role. I did not know Yuchi Jian before. Professor Chen Fu forwarded this book to me. I read it carefully and found it very useful.

First, this book is easy to understand. Book is written to be read. If it is hard to understand, nobody will read it. Except for the small number of specialists, most of us are more used to master knowledge and accept ideas in a relaxed state. To me, the essence of this book is to explore how the Chinese economy, especially the manufacturing industry can transform from big to strong. This means to be a serious topic. A Ph.D. student may utilize various theories (e.g. mathematical economics, advanced macroeconomics, advanced microeconomics) and draw conclusions through complex arguments. By contrast, this book has been written in a way that everyone accepts the new idea of "asset-light" in a relaxed state. Following this idea, the Chinese economy can be transformed and upgraded. The gap between China and advanced countries has been bridged, finally fulfilling the "China Dream" (i.e. the rejuvenation of the Chinese nation). As far as I can see, "Big Data" written by Tu Zipei (whom I did not know either) is a good book on big data. This book written by Yuchi Jian is a very good book that integrates Internet and big data with real economy. Although contents are on Internet and industrial economy, the language used is so elegant.

Second, this book is full of thoughts. Looking back the human history, most books that are past down to future generations are full of thoughts. This is an era of knowledge explosion. However, it deserves deep thinking what can be left after the explosion. Everyday, a magnanimity of books are published and then converted to pulp straightaway. In essence, it is because of the lack of thoughts. After carefully reading, I believe there are plenty of thoughts in this book. For instance, the Germany "Industry 4.0" is summarized to a transformation from "asset-heavy" industrial development mode to "asset-light" industrial Internet development mode with an aim to operate the core values of the industrial ecological chain. The Chinese leading enterprises should adopt a "stripped" strategy for which their large number of production, processing and assembling workshops and service links should be gradually passed onto SMEs. By contrast, leading enterprises should pay more attention to the R&D of core technologies, high-end manufacturing, establishing industrial standards and digital standards, as well as the deep integration between the Internet, software, big data and informationization. Ultimately, the Chinese leading enterprises must adopt the new development mode, i.e. "asset-light" industrial Internet.

Third, this book is full of convincing cases. The author has conducted thorough study of the industrial economy of various developed countries such as Germany, USA and Japan. The use of cases has made it very easy for readers to understand the principles and theories.

Of course, no product is perfect. I trust the author will continuously improve this book. I am looking forward to the future editions of this book.

April 2016

Zhenhua Zhao
Economist, the Dean of the Graduate
College of the Party School of the
Central Committee of CPC
Beijing, China

Foreword by Yiming Wang

I was invited to write recommendation forward for this book. Therefore, I personally met Yuchi Jian, the author of this book. Despite a short meeting, I am very impressed by his thoughts on the Internet Economy.

During the travel, I managed to find time to read this book. To be frank, nowadays very few books are so interesting that readers are happy to read from the beginning to the end. However, I read this book all at once. The topics covered in this book is very interesting. In addition, the logic is very clear, and the art of writing is shown throughout the book. Internet technologies may seem very hard to understand, however this book is not hard to follow. More importantly, the analysis and argument in this book clearly show the unique opinion and in-depth thinking of the author.

Since its introduction, there is never-ending debate of Internet. However, in most cases, Internet is considered as a technology, tool or approach. By contrast, very few studies approach Internet in terms of economic form. From the perspective of industrial development mode and production mode, this book proposed that "asset-heavy" mode which aims to operate the value-added of the enterprise itself, will be replaced by "asset-light" development mode which aims to operate the core values of the industrial ecological chain. In terms of economic form, the traditional manufacturing industry, financial investment industry and service industry are relatively independent. By contrast, following the new development mode of Internet Economy, the Internet and Industrial Internet will become the core of the national economy. Meanwhile, the Internet and Industrial Internet will be deeply integrated with traditional manufacturing industry, financial investment industry and service industry. Indeed, the boundary of industries becomes increasingly blurred. Both Germany Industry 4.0 and U.S. Internet companies (presented by Google and Apple) will result in revolution of the traditional industries. Internet economy represents the direction of industrial development. Indeed, the core values of industries are flowing to the Internet industry. Traditional manufacturing industry, financial investment industry and service industry will be completely transformed by the Internet industry.

How to keep the pace of the Internet economy? According to the author, there are two fundamental approaches for the transformation and upgrading of the Chinese traditional industries to the Internet industry. These two approaches are: "+ Internet" model, i.e. letting software company and Internet company "be married" to manufacturing enterprises; and "Internet +" model, i.e. letting manufacturing enterprise "be married" to software company and Internet company. This is undoubtedly a key direction to adjust the national industrial structure in China. Both models will lead to the reform of traditional industry and economic form. In this book, the author has used a large number of typical cases to illustrate the reasoning that core values are flowing to Internet industry, software industry and high-end manufacturing industry. These are very convincing.

I highly agree with the author that priority should be given to the impacts of Internet on the traditional manufacturing industry and service industry in China and beyond. The author pointed out that leading enterprises should adopt a "stripped" strategy that those non-core technologies and businesses could be handed to specialized SMEs. By contrast, the leading enterprises should pay more attention to the R&D of core technologies, high-end manufacturing, establishing industrial standards and digital standards, as well as the deep integration between the Internet, software, big data and informationization. They need to place more focuses on developing the industrial technology systems with industrial supply chains and service chains. Based on the allocation of superior resources at the national and international level, new industrial ecosystems will be formed. Those leading enterprises should be upgraded to "asset-light" industrial Internet companies with an aim to operate core values of the industrial ecology. They will stay at the top of industrial value chain, keeping up with the new Internet economy. Indeed, these thoughts and opinions are very forward-looking.

Of course, I do not necessarily agree with all opinions proposed by the author. For instance, the author proposed three strategic base points, i.e. planned economy, market economy and Internet economy. Consequently, a new economic institution, structure and mode can be developed for China, i.e. "e^3 theory". The author might have mixed up two things, i.e. planned economy and the effectiveness of government intervention. As a kind of institutional structure, it is difficult to integrate planned economy with market economy. The key to promote the socialist market system is the decisive role of market in resource allocation with higher effectiveness of government intervention, from the perspective of institutional form. Surely, there will be new demand for the institutional reform of the national economy due to the development of Internet economy. However, it is not appropriate to mix the instructional form with economic form.

Of course, I never deny the core opinions and unique value of this book. This book is absolutely worth reading, no matter you are entrepreneur that are keen to explore the Internet industry, or you are specialist that are interested in the research of Internet industry and Internet economy.

Therefore, I strongly recommend this book to you.

April 2016

Yiming Wang
Economist, Deputy Director-General
of the Economic Development
Research Center of the State Council
Beijing, China

Foreword by Nanning Zheng

I have known Yuchi Jian since 2000. Due to working commitments, both of us participated into discussions on the strategic research and planning related to manufacturing and information technology. At that time, there is a new tide of Internet development. During that period of time, Yuchi Jian is very keen to discuss from experts on the future development of information technology, impacts of continuous innovation of information technology and its integration with industries on the economy. He is very good at independent thinking, often coming up with unique opinions on the technological innovation as well as the reform of economic development mode.

As soon as reading the table of contents, I have been impressed and willing to finish the reading all at once. The opinions, arguments and examples are all closely related to the national strategy on Internet Economy. It clearly pointed out the key for the transformation of the Chinese economy and industrial development mode. I believe this book will be very popular. Indeed, I strongly recommend entrepreneurs, industry circles, financial investment circles, economic circles and government decision-makers to read this book and participate into the debate on Internet Economy and the new development mode.

Recently, the impacts of Internet on economy, industries and enterprises have drawn extensive attention. This is especially the case for the traditional manufacturing industry, financial investment industry and service industry. As a kind of information technology, how Internet can make such profound impacts on economy, industries and enterprises? I am sure everyone is thinking about this question. Based on the extensive research, the author proposed that Internet has become a new economic theory rather than a core technology. This has helped us to gain a better understanding of the major transformation derived from Internet on the traditional manufacturing industry, financial investment industry and service industry. It seems that there is continuous innovation of information technology that is represented by Internet, software and big data. The emerging new production forces will inevitably lead to the adjustment of production relations accordingly. Indeed, opportunities and challenges co-exist.

Today, most people perceive that the aim of Germany's Industry 4.0 is to achieve smart manufacturing. However, the author has reminded us that it is not that simple. In fact, Industry 4.0 is closely related to Internet industry, industrial Internet industry as well as the new industrial development mode. In essence, Industry 4.0 is designed to facilitate the transformation of once-powerful Germany manufacturing industry to transform from "asset-heavy" to "asset-light" development mode, as a response to enormous pressure from the new Internet economy development mode implemented by the United States. I agree with this argument. At the national level, by implementing Industry 4.0, Germany can continue to occupy the top of industries together with other developed countries such as the United States in terms of energy conservation and emission reduction. They can continue to influence the quality and speed of economic development in developing countries such as China and India. At the enterprise level, by implementing Industry 4.0, Germany companies can move towards the top level of industrial value chain and keep the advantageous position. Therefore, Industry 4.0 is indeed a matter related to Internet industry, industrial Internet industry as well as the new industrial development mode.

Recent years also witnessed rich discussions on the Third Industry Revolution initiated by the United States. The whole process is extensively reviewed in this book. The Western developed countries and their leading enterprises proposed the globalization of economy since 1990s, followed by the outsourcing of manufacturing and service activities. With the support of informationization, Internet companies (represented by Google and Apple, two U.S. companies) emerged, which resulted in the changes of industrial development mode. This is the very first time that origin, form and impacts of Internet and industrial Internet were reviewed in a systematic manner. "+ Internet" model and "Internet +" models were discussed, so as their impacts on the traditional manufacturing industry, financial investment industry and service industry. Consequently, the economy and industries are completely transformed. Finally this book proposed the fundamentals and cores of the Third Industrial Revolution. All these deserve in-depth study and discussions.

In this book, the author has extensively discussed the critical role of industrial standards and their digital standards played on various industries such as software, Internet, big data, manufacturing, financial investment and services. Indeed, in the era of Internet Economy, digital standards is vital for the concept of modern industries, industrial ecology and big data. No digital standards, no future of the manufacturing industry. No digital standards, no grand development of Internet industry and industrial Internet industry. To keep pace with the Internet economy, it is necessary to start with fundamentals such as digital standards.

Internet has significant impacts on economy and industries because it provides values derived from continuous innovation of businesses, management and services. Products of today and future must feature with digitization, network orientation and smart. The industrial development mode has to be compatible with the Internet economy. Core values will flow to industries such as Internet, software, chips, content services and high-end manufacturing. In this book, core values of

Internet economy are clearly proposed, which deserves thorough thinking and discussion.

I am very pleased to see that the author has pointed out the root causes of structural issues associated with the Chinese economy. He identified the structural issues associated with the Chinese economy, and concluded the appropriate structure in the future. These are based on the thorough analysis of Industry 4.0 (Germany) and the Third Industrial Revolution (USA), as well as the impacts of planned economy mechanism on the development of Chinese economy in last decades. As a consequence, the author proposed three strategic base points in Chap. 8. Subsequently, he proposed the "e^3 economic theory", which is wonderful. Meanwhile, in Chap. 9, the author clearly stated "Four-step" strategy for the transformation and upgrading of Chinese industries and enterprises. This is of practical significance. For instance, it is recommended that the leading enterprises should follow the "asset-light" Industrial Internet development mode. In my opinion, this also deserves thorough study and discussion.

Internet economy is fast approaching. As a result, the traditional manufacturing industry, financial investment industry and service industry that we are familiar with, will be changed significantly. We have just started the study and discussion of Internet economy. I recommend this book to encourage the participation into the discussions of Internet economy. As a result, China will become even stronger.

December 2015

Nanning Zheng
Academician of the Academy of Engineering of China, Chairman of the Chinese Association of Automation, Former President of the Xi'an Jiaotong University
Beijing, China

Acknowledgements

I have been thinking about writing this book three years ago. In fact, this book is also a summary of my personal work for many years. When starting to write in mid-May 2015, I was not happy with my research as some key issues were not studied in detail and in depth. Therefore, I was sometime confused and upset. I would like to thank Mr. Huang Yongyou and Prof. Liu Feng for their encouragement and help. Constantly exerting a lot of pressures on me, they had high expectations, e.g. far-reaching vision, thorough analysis and well-understood cases. These are expected to make contributions to China to advance into the Internet economy. In the process of writing, I often experienced challenges and difficulties. Every time I was depressed, I held expectation that I could be born again.

I am very grateful to two economists. They are: Zhao Zhenhua, the Dean of the Graduate College of the Party School of the Central Committee of CPC and Wang Yiming, the deputy director-general of the Economic Development Research Center of the State Council. Both of them are keeping busy all the time, and I was not familiar with them. In their busy schedule, they reviewed my manuscript and wrote recommendations. Similarly, they pointed out shortcomings as well as the direction of improvement. These are of great support and encouragement to me. It is not necessary to have met with each other to become the best friends, the key lies in the communication of spirit and ideas.

I owe great thanks to Academician Zheng Nanning for writing recommendation for this book. Once reading the table of contents, Academician Zheng believes that this is a good book that fits the current situation of national economic development. He believes that this book would be popular and shed significant social impact. This provides great encouragement and support to me. I would like to thank Academician Mei Hong, Academician Liao Xiangke, Mr. Gao Xinmin (Executive Vice Chairman of the Internet Society of China), Mr. Wang Wenjing (Chairman & CEO of Yonyou Network Co Ltd.), and Mr. Wang Xingshan (Chairman and CEO of Inspur Technologies Co. Ltd.) for writing recommendations for this book. They provided significant encouragement, trust and a number of invaluable suggestions. They are all elites and pillars of science and technology, industry and economy circles in China, guiding my work and life. Thank you once again. I would like to express my special thanks to Academician Liao Xiangke, who read my book in the midst of pressing affairs. While affirming and praising, he also made very specific suggestions for revision, which has benefited me a lot. It is my honor to know him. He is indeed the pride of Hunan people and the pride of the Chinese science and technology circle.

I would also like to thank Mr. "No", Liu Peng, for his support and help. It is not easy at all for my book to be recognized by him, because he said "No" to everyone else. Now, he praised my deep thinking of the Internet. I would also like to thank Prof. Wang Jianmin, Liu Yingbo, Ding Xiangqian and Dai Guojun for their encouragement and support. They also put forward very useful suggestions so that I am confident in writing a good book.

I would like to express all my thanks to Mr. Zuo Jian. He made great and standout efforts, and used his wisdom to translate my book in English. We are twins each other, he is the book in English and I am the book in Chinese. The essence of translation has always been an arduous and re-creation work, for which we have shared the same heart and soul. What a great cooperation and contribution!

I would like to express special thanks to Mr. Kris Singh, the CEO of Service Research and Innovation Institute (SRII) of the USA. He encouraged me to translate my book into the English version. This will facilitate the introduction of my e^3 economic theory as well as the integrated development (i.e. Internet, big data and AI with the industries in China) as soon as possible to the United States of America and the world. Mr. Kris Singh and I met and discussed in the Tsinghua University, Beijing. We both soon found that we have shared the same industrial and economic values as well as the same understandings of innovation. We now become friends of life.

I would like to express my special thanks to Mr. Zhang Zimao, the former President of the Beijing Jiaotong University Press. He has guided me with his profound knowledge, strategic courage and unique mind. He has discussed key chapters and sections with me for a long period of time and encouraged me to write a valuable book which will benefit the nation and the society. I would also like to thank Mr. Zhang Zimao's publishing team for their professional and meticulous guidance. It is impossible to get this book published without their assistance and collaboration. Thank you again.

I am very grateful to Ms. Liu Yan. As my colleague, she spent her valuable time and energy in reading my manuscript and providing a lot of valuable suggestions. She raised her personal concerns about my description of the state-owned and central enterprises in the book. I personally think that the vast majority of entrepreneurs in state-owned and central enterprises love the country, people and society. They are working very hard thus should not be blamed excessively. I simply made a general summary or evaluation of challenges and drawbacks related to the system, mechanism, management and services in those enterprises. The main aim is to encourage the state-owned enterprises to develop the new economic development model so that the transformation and upgrading can be achieved as soon as possible. As a consequence, they can become bigger and stronger.

I would like to thank three female classmates: Lin Qiong, Sun Jing, Huang Xiaoyan (and her father), and my colleagues: Wen Bin, Yang jiankun, Sun Quan, Li Honggang, Lin Shenwen, Chen Hao, Shang Cuiting, Cui Chenyue, Wang Chen, Bai Fanlu. Thank you very much for your selfless and enthusiastic help during my writing, which makes me feel warm-hearted.

I would like to thank my family for their great understanding and support to my writing. Since mid-May 2015, it has been very rare that I could accompany them to have family dinner or enjoy outdoor activities during weekends. Thank you again.

Beijing, China
May 2019

Jian Yuchi

Recommendations

The Internet economy is fast approaching, however its essence remains unclear. Based on extensive analysis of the Industry 4.0 (Germany) and the Third Industrial Revolution (USA), the author boldly proposed the "asset-light" industrial Internet development mode that controls the industrial standards, data and software with an aim of operating core values of industrial ecological chain. He pointed out two different models, i.e. "+ Internet" model that is adopted by traditional manufacturing industry and "Internet +" model that is adopted by the Internet industry. Despite with different starting points, these two models are oriented to the same direction, i.e. the industrial Internet business. I recommend this book because it helps to initiate the in-depth thinking of essence of Internet. With a better understanding of Internet economy, it is more likely to achieve success for the Internet economy in China.

January 2016

Liao Xiangke
Academician of the Academy
of Engineering of China

The new generation of information technology, represented by Internet, has substantially changed the human society, or even likely re-organize the human society. We are heading toward the era of Informationization 3.0 (i.e. Smart era) that is featured with deep data mining and application. In this book, Internet was approached from the perspective of economic development mode. The new economic theory of "Industrial Internet" was discussed, which is unique and worth reading.

December 2015

Mei Hong
Academician of the Academy of Science
of China, Group Leader for the Information
Technology Strand of the 863 Program

Two key trends of enterprise development are Internet orientation and finance orientation. It is likely that all enterprises will become Internet company (driven by data) and financial firm.

Mr. Yuchi Jian has critically analyzed enterprises, manufacturing industry, industries and even the overall economy from the perspectives of technology and economy. This book is worth reading by both academics and industry practitioners.

January 2016

Wang Wenjing
Chairman and CEO of the Yonyou
Network Technology Co., Ltd.

Nowadays, Internet is fast approaching which re-shapes the current industry's form. Every enterprise is re-considering its value as well as positioning in the industry ecology. It is inevitable that enterprises will achieve transformation and upgrading with the assistance of a software platform which is Internet-oriented. The integration with Internet will be accelerated. Meanwhile, the new generation software platform that is driven by big data will provide key support in this process. In this book, the impacts of Internet on industries and the inherent core values were deeply analyzed. A unique lens was used to show the blueprint of industrial ecology as well as the pathway of transformation and upgrading. This book is worth reading. I am very keen to recommend this book to entrepreneurs.

January 2016

Wang Xingshan
Executive Director of the Inspur Group Co., Ltd.

It is inevitable that Internet economy will change the world which we are familiar with. We have just started the discussion of Internet economy. I recommend this book to facilitate the studying and discussion of Internet economy. Let us make this world an even better place.

December 2015

Gao Xinmin
Executive Vice Chairman, the Internet Society of China

Contents

About the Author

Prof. Jian Yuchi born in 1963, have qualifications in computer, public administration and Internet economy. His book, Value Proposition (Chinese version: 价值魔方) is one of the bestsellers with tens of thousands of readers in China, and was honored with the first prize on China's enterprise reform and development by the state in 2018. He is a senior researcher of the China enterprise reform and development research association which is one of the top think tanks in China's economic reform and development. He has been invited to give speeches more than 150 times by the main industries and enterprises, Shanghai Stock Exchange, universities, R&D institutions, industrial development zones and science parks, high-level government agencies and major events across China. He has been well recognized as a prominent economist in China in the era of new Internet and big data economy. He proposed "e^3" economic development theory, which has played a crucial role in the integrated development of Internet, Industrial Internet, Big Data, AI and manufacturing industry in China.

Chapter 1
Relationship Among Computer, Software, Internet and Industries in Last Eight Decades: Industrial Revolution

It is imperative to understand the integration of industry with information technology following the development route of computer, software and Internet. The close relationship between industry and information technology during the last eight decades can be classified into four stages. Consequently, a better understanding of Industry 4.0 can be achieved.

In 1930s, the Von Neumann architecture was proposed for computers. In next five decades, the take-up of computers in the industry is very slow. In most cases, computers were used to perform the specialized scientific calculation. During that era, computers are very rare and not accessible to ordinary people. Using assembly language, only a small number of professionals could use computers. This era is called Industry 1.0.

Let's take the Boeing aircraft as an example. The major overlap between aircraft and computer is the fluid mechanics. Fluid dynamics plays a crucial role in analyzing the aerodynamic shape, structure and weight of aircraft. Similarly, fluid dynamics helps to determine the relationship between these factors and other key parameters of flight (e.g. resistance force, lifting force and safety). These findings are useful to optimize the aircraft design. These processes are still current in modern aircraft manufacturing industry. In past, aircraft design relies on manual approach through the drawing boards and pencils. Each aircraft consists of hundreds and thousands of parts and components. Therefore, thousands of drawings have to be developed by a large number of specialists. These drawings are heavy and take spaces for documentation. Any change to the original design has to start with locating the corresponding drawings from the cabinet. Such process was very inefficient. Later, calculators with relatively simple structure were invented and used to assist the computing. In Boeing Company, manual documenting was the main approach to capture various production plans and enterprise management plan. As a consequence, a large scale of documents were produced. Similarly, a large number of professionals are required for the maintenance and services of aircraft. Therefore, the shape, structure and functions of aircraft were designed as simple as possible during that era. The design and manufacturing of an aircraft took a long period of time and required inputs from a large

J. Yuchi, *Value Proposition*,
https://doi.org/10.1007/978-981-15-5025-6_1

number of professionals. Similarly, it demands massive investment from enterprises and the government.

The 1980s marked the Industry 2.0 era, namely the information age. This era witnessed the joint innovation between Microsoft and Intel on computer, software technology and product, and industrial development modes. Consequently, the manufacturers continuously launch personal computers, commercial computers and servers with "Win-tel" architecture. Due to advantages such as high performance and low prices, these computers and servers took a large market share. Meanwhile, high-level programming language such as BASIC, FORTRAN, C emerged which were simple and easy to learn. As a result, during the production activities, everyone could use the computer to handle daily business and data processing. Subsequently, the software is separated from the computer products and its industry. This indicates the emerging of the software products and industry, independent of the computer industry. Later, computer-aided software has been widely used in various production stages such as: R&D, design, processing, manufacturing, management and servicing. Through innovation, the rapid development of computer industry also produced a large number of industrial software companies. The software industry is gradually integrated with the manufacturing industry. For instance, excellent software vendors include: (1) for the design of mechanical products: CATIA (Dassault, France) and UG (acquired by Siemens, Germany); (2) for the enterprise resource management: Oracle (USA), SAP (Germany) as well as Kingdee, UF and Inspur (China); For the enterprise manufacturing and technologies, there were relatively personalized manufacturing execution systems such as MES as well as the process software which were applied to the CNC machine tools, industrial robots and automation equipment; For the services, PLM was designed for the life-cycle management of products, and MRO was designed to assist the operation and maintenance, monitoring and service of complex mechanical products. These software were continuously upgraded and always faced fierce competition. Since then, efforts were made to connect all above mentioned processes and their applications so that an Intranet is formed. Ultimately CIMS(computer integrated manufacturing system) is developed within the enterprise. Industry 2.0 era witness the emergence of a large number of core software and industrial software oriented to the manufacturing industry. The industrial software industry has gained rapid growth, which provides necessary support to the development of industrial and manufacturing enterprises. As a result, the duration for product's R&D cycle is significantly reduced and the efficiency of enterprise management is greatly enhanced. Other benefits include: significant reduction of manufacturing and processing costs, significant improvement of the product service, significant increase of the enterprise benefits and profits, and significant reduction of the labor intensity of technical staff and workers. In other words, the manufacturing industry entered a new era due to the innovation of computer, software and network. Since then, computer, software, network and other information technologies have been deeply integrated with the industry.

In Industry 2.0 era, the Boeing Company has achieved digital management of design drawings. Meanwhile, a large number of data, experience and knowledge are accumulated during the production practices over many years. Such information has been integrated into the aircraft design and simulation software. This has greatly shortened the aircraft's R&D and design cycle. New software were developed continuously and applied in manufacturing process. In addition, the enterprise resource management software was utilized to integrate and deploy various production and service resources. These measures have led to a number of benefits such as: the continuous reduction of cost, continuous improvement of efficiency and labor productivity. Finally, the Boeing Company has realized the full integration of design, manufacturing, management and services. As a result, Boeing's products (e.g. 737, 757, 767, 777) were quickly launched with strong competitiveness advantage. The Boeing Company maintained a global leadership position.

It can be observed that those key enterprises from USA and other western developed countries were always in the leading position throughout the Industry 2.0 era. This is due to their solid foundation, adequate funds, advanced technologies and continuous innovation. Consequently, these enterprises gained rapid globalization in this period. This is followed by a large number of "outsourcing" activities related to processing, manufacturing and services. Most of enterprises in developing countries (including China) are still in the development stage of Industry 2.0 era.

Since 1990s, the Internet has been widely adopted. This marks the beginning of Industry 3.0 era, namely the Internet era. The fundamental concept of Internet is the world wide web (WWW). WWW refers to the web which can be globally implemented, operated and interacted with a high cost-effectiveness. Such explanation can be further simplified that WWW is a high performing, fast and global optical transmission network which everyone can use and afford. Internet can connect computers and terminals over the world efficiently and effectively to achieve data communication. Meanwhile, WWW can integrate the resources, business and data of the global manufacturing industry as well as its associated services industry through software applications, leading to a new industrial ecosystem. Consequently, the Internet can transform the massive amount of data generated within this new ecosystem into new knowledge. The associated benefits include: smart design and manufacturing process, the global collaboration and optimized allocation of resources, the improved management efficiency, and customized and life-cycle-oriented services. Finally, the investment return can be maximized and the associated risks can be minimized.

Because of accumulated experience in Industry 2.0 era, the leading manufacturers of developed countries (e.g. USA, Japan and EU) have carried out the Internet-based global restructure of their systems, businesses, management and services, which is achieved via the development of Internet, software and data centers. As a result, digitization and industrialization are highly integrated. This has strongly supported and accelerated the globalization. In fact, these countries and companies have accumulated capacities on implementing Internet-oriented globalization, management and services. This laid solid foundation for the emergence of big data and artificial intelligence (AI) services. In the Internet era, the core value of big data and AI is widely recognized. Meanwhile, software companies re-developed their products to be Internet-oriented. This is to meet the market demands so that the enterprise can sustain development. Indeed, all software products and associated services are Internet-oriented.

It can be observed that the Boeing Company has laid solid foundation in Industry 2.0 era. In the Industry 3.0 era, Boeing further unified its digital standards and restructured its business, management and service following an Internet-oriented approach. The design, manufacturing, management and services of Boeing 787 is achieved via the collaboration among various enterprises from 48 countries. Internet is widely adopted for the operation management and services of Boeing Company. More focuses are placed on fulfilling demands of customers (airlines) as well as the life-cycle services of aircraft. The original business model of Boeing Company was simply selling aircraft and parts. It is now transformed to a new business model featured with the integration of selling aircraft with life-cycle services. Such business model helps to acquire the values across the life-cycle of every single Boeing aircraft. In simple terms, there is a revolution of the business model of Boeing Company which was transformed from product-oriented to service-oriented. More attention is paid to the satisfaction of passengers on a range of issues in the cabin such as: storage of luggage, vision, lighting, temperature, humidity, oxygen supply, comfort, network services, entertainment, bathroom, etc. Boeing Company's concept of design, management, manufacturing and service is cutting edge around the world. Indeed, Boeing Company is the leader for the transformation of global manufacturing industry.

In the Industry 3.0 era, a number of new technologies emerged such as: smartphone, mobile terminal, 3G and 4G network communication. Consequently, the Internet companies, e.g. Google and Amazon (USA), and Alibaba, Tencent and Baidu (China), found the niche market in the service industry. Through innovations of the App software, service model and service experience, these Internet companies let those traditional service industries and their enterprises be experiencing significant revolution. In the future, the Internet, 5G, big data and AI will also show significant impacts on the traditional manufacturing, financial investment and even software industries.

It is worth noting that most of leading Chinese manufacturers are not into the Industry 3.0 era, namely the Internet era. A lot of efforts are still required.

It needs to be cleared off and emphasized that there is fundamental difference for enterprise's value and development strategy between the information age and the Internet era. For the information age or stage, enterprise places focus on products where the business model is featured with the operation of products and enterprise. It is very rare that product is sold below its cost. Various software such as the product design (CAD), enterprise resource management (ERP), manufacturing execution system (MES) emerged as production tools that aim to improve the efficiency and reduce the costs. This is a typical form of traditional manufacturing industry. By contrast, for the Internet era, the service is the key focus. In other words, the objective of operation for leading enterprises is the industrial ecosystem rather than product. Specifically, the goal and the main business model of an enterprise is to operate various core values in its industrial ecosystem. Such industrial ecosystem covers not only the form of manufacturing industry, but also the form of industries such as software, chips, services, Internet, big data, AI, and even financial investment. An industrial ecosystem may involve the cooperation and collaboration among hundreds and thousands of enterprises. This is a new-type of complex business form, which is often called the Internet industry or industrial Internet business today. As such enterprise operates various core values in various industries, as a result, it is not unusual for its operation and business model that some products and services are provided to customers at a lowest price and even free of charge. It is not hard to imagine that such lower-price or "free" products and services provided by Internet based enterprises will have significant competitiveness advantage compared to traditional product-oriented enterprises, and even lead to a revolution to traditional product-oriented enterprises in the long run. Apparently, such novel production and business model is not possible without the support from software, Internet, big data and even AI. In such a industrial ecosystem featured with the collaborative development of multiple industries, enterprises and agents, the most crucial factor is to achieve the free and efficient information exchange among entities (e.g. people to people, people to things and things to things, and enterprise to enterprise). Therefore, it is imperative within such an industrial ecosystem to define the universal language and rules, i.e. universal standards. This is the key for leading enterprises to achieve success. During this process, data, big data and their analysis (i.e. AI) will inevitably March into the central platform of industry, and become the core value of industry.

Since 2010, some European experts and scholars began to talk about future goals and impact of new technologies such as: Internet of Things, 5G, robotics and artificial intelligence. However, there was lack of common understanding of whether that meant the emergence of the next round of the industrial revolution (i.e. the Industry 4.0). For a long period of time, the impact of software, Internet, big data, artificial intelligence and other information technologies were assessed from the technical perspective. The rapid development of information technology makes it difficult to follow even by professionals. This has resulted in confusion to entrepreneurs, government officials, industry practitioners, economists and financial investors. A number of questions remain. What is the end of information technology development? Is it worthwhile following the development of information technology? What is the appropriate view to the industrial evolution based on Internet and big data? Are Internet and Big Data simply technical terms? What are actions to be taken by entrepreneurs, investors and government officials? It is imperative to rethink the fundamentals and nature of information technology development. I think that it is more appropriate to focus on the impacts of Internet and its development on the new production relations and new production mode from the economic and industrial perspective. This will be more beneficial than just focusing on the achievement of these new technologies. In other words, it will be of more practical significance to discuss and investigate how the information technology (i.e. Internet, software, cloud-computing, big data and AI) can be better incorporated into real economy and new industrial ecology. Therefore, the future direction will be the deep integration among various industries such as: the manufacturing, service and financial investment, and software, Internet, 5G, big data, and AI technologies. Indeed, there are both opportunities and challenges associated with a new industrial development model featured with "deindustrialization" and "light-asset", namely the Industry 4.0 era.

On the basis of Industry 3.0, those leading manufacturers such as: Boeing and GE (USA), the Rolls Royce (UK), Siemens (Germany) continuously employ software and Internet in their business activities. This has led to the emergence of the "+ Internet" model. Meanwhile, the Internet companies and software companies (e.g. Google, Apple and Microsoft) have started to explore the traditional market of manufacturing industry or real economy such as: automotive, robotics, health, energy and transportation. They attempted to develop industry-oriented Internet, resulting in the emergence of "Internet +" model. In recent years, the Apple Inc. has adopted the open and collaborative innovations in its design, manufacturing, management and service of smart phone and mobile terminal. Especially, the Apple Company focuses on the core values of its industrial ecosystem where a new industrial Internet model of "deindustrialization" and "light-assets" is adopted. This has facilitated better understanding of the industry 4.0, industry-oriented Internet or the industrial Internet.

Many people may have a same question. Why Steve Jobs is so popular? Is it due to the fashion and industrial design, or App software and service experience? Essentially, Steve Jobs led the transformation of the industrial development mode of the manufacturing industry which we are accustomed to for more than three centuries. In other words, Steve Jobs changed the production relations, production mode and industrial development model of the traditional manufacturing industry. This is based on a deep understanding of the industrial software, as well as creatively applying the industrial standards, Internet, big data and AI. He in fact created a new industrial ecosystem. In the traditional manufacturing industry, manufacturers generally have to design, process, manufacture and assemble their products by themselves if they want to sell products, create values and gain profits. In essence, product is the core of traditional manufacturing model, which can be simplified as the "direct production mode". However, the smart-phones and other products sold by Apple Inc. are service-centered, which are manufactured by other enterprises within its industrial ecosystem. The Apple Inc. is not involved in the manufacturing of parts, even the final assembly and testing of the finished products. This is so-called "zero manufacturing". In other words, the Apple Inc. does not have to invest massive capital and resources (e.g. lands, plants, equipment and manpower) in manufacturing, assembly and testing processes. Instead, Apple Inc. outsources such enormous and complex tasks to a large number of specialized enterprises within its industrial ecosystem. Of course, the profits generated from the manufacturing activities are also shared within Apple's industrial ecosystem. This new mode can be called "indirect production" or "distributed-networked-production" mode. During the complex manufacturing process, Apple Inc. in fact avoids the potential investment and operational risks derived from the massive "heavy assets" investment by means of sharing profits with other cooperative manufacturing enterprises within its industrial ecosystem. This new mode of the industrial development is so called the innovative "deindustrialization" and "light-asset" mode. However, challenges exist to deliver Apple products in a high-quality, within the timeframe and within the budget according to the production and sales planning. The issue associated with any production activity will have certain implication, even on Apple's reputation. As most of production activities are outsourced, challenges exist to effectively organize, control and manage those tasks. Indeed, compared to the traditional "direct manufacturing" mode, significant challenge exists for Apple Inc. to accomplish technology execution and organizational control. Faced with its massive industrial supply chain, service chain and ecological chain, various industrial software should be first developed by Apple Inc. and be then incorporated into all enterprises within the entire industrial ecosystem. This helps to achieve the precise management and control based on standards, Internet, big data and AI during various processes. Once achieved, an industrial technology threshold is set for the entire industry which is hard to overpass. This mode is indeed a fundamental revolution to the traditional manufacturing industry, which facilitates a better understanding of the industrial software, Internet, cloud-computing, big data, AI and standards as well as their impacts on the manufacturing industry. The deep application of information technology in manufacturing industry can not only improve the productivity, but also change the production relations, production mode and industrial development model. This results in the significant improvement of enterprise's efficiency and systematic reduction of costs and investment risks. This revolutionary change also suggests that the value-added and innovations of the future manufacturing industry has been quietly transferred to the development of information technology (e.g. the standards, software, Internet, big data and AI). This will become the future direction of industry. Therefore, the manufacturing industry tomorrow will no longer be what it is today.

You might question why Apple Inc. chooses such complex production mode, production relations, and industrial development model instead of the traditional production mode and industrial development model? What are the benefits and advantages of doing so? In my opinion, such new approach and development model help to achieve the best and optimal state from the economics perspective.

First, the profits derived from the investment and operation can be maximized. Apple Inc. makes most of the market mechanism in its industrial ecosystem. This has facilitated the full competition among enterprises in the industrial ecosystem. Each enterprise has to demonstrate the lowest cost, best performance and quality and best service of their products. As a consequence, Apple Inc. took over 90% of profit from its smart-phone products. By contrast, those enterprises in China which provide processing, manufacturing and assembly services for Apple Inc. only shared 4.6% of profit in spite of investing massive amount of resources in plants, equipment and workers.

Second, the investment and operation risk is minimized, and the investment exit mechanism is optimized. This is especially important as it is not guaranteed that the investment is always associated with desired return and profit. In fact, an economic methodology has been sought and required all the time by entrepreneurs and investors to better solve this problem. According to Apple's industrial development model (i.e. light-asset), a company can simply and quickly announce to exit the market, if its product is determined as oversupplied due to the excessive market competition. It can be observed that Apple Inc. can exit the market with even "zero loss" and "zero risk" due to the "light-asset" model adopted in the massive manufacturing process. On the contrary, a large number of OEM enterprises bear the risks of overcapacity and heavy investment for Apple Inc. These enterprises provide processing, manufacturing and assembly services to Apple Inc.

In developing countries such as China and India, a large number of OEM enterprises are located in the downstream of supply chain. Providing services such as processing, manufacturing and assembly, these enterprises will inevitably deal with the overcapacity and investment risks transferred by western companies. In recent years, the problem of over-capacity in China has drawn a growing level of concern, which can be regarded as the economic crisis of an industry in broad sense. When plants and equipment are left idle, it is a hard decision whether or not making redundancies. When it becomes more serious, such issue can be developed into a social problem. In fact, western economists have always been discussing the repeated occurrence of the product oversupply due to the excessive competition. This is arguably due to the greedy nature of human beings and investors. As a consequence, there are cyclical occurrence of the overcapacity and economic crisis. Therefore, economists, as well as investors and entrepreneurs, have been looking for ways to avoid and solve the problem of overcapacity and economic crisis. Apple's new production relations, production mode and industrial development model are based on standards, software, Internet, big data and AI. Is this the methodology to deal with overcapacity and economic crisis? Should Chinese enterprises follow this "light-asset" approach?

Third, for a long period of time, the financial investment industry generally adopts the traditional approaches (e.g. staff examination, debriefing, financial auditing) to control the risk associated with large-scale loans or investment. However, these approaches are not sustainable and will inevitably lead to higher level of risks. Therefore, the financial investment industry always has a dream in which they can obtain the operation and management data of borrowers. Through data transparency of both parties, investment risks can be accurately controlled. We might recall the industrial ecosystem which Apple Inc. develops and controls through the standards, software, Internet, and big data. Through a massive industrial ecosystem consisting of tens of thousands of enterprises, Apple Inc. in fact quietly controls all sets of data related to the design, manufacturing, management, service and operation when cooperating with other enterprises. This is a typical form of big data for an industry. All relevant data within the industrial ecosystem are transparent to the Apple Inc. & "Apple Cloud". The big data will inevitably produce endless core values. Apple Inc. may carry out the big data analysis to evaluate which are the competitive and high-quality enterprises within its industrial ecosystem, and then may in turn carry out investments in such enterprises. As a result, the investment efficiency will be maximized, the investment risks can be minimized, and the investment security can be optimized. Consequently, the investment is more likely to be successful. Therefore, it is inevitable that Apple Inc. transforms to a financial investment service company in its industry. Those traditional financial investment companies and banks may become partners of Apple Inc. A new financial investment model emerged, which is industry-oriented and big data-AI-based for the entire industrial ecosystem. In other words, companies like Apple Inc. will gradually replace and even revolutionize the traditional financial investment companies and banks. In my opinion, such a new financial investment mode based on the big data and AI can be called the third generation of financial industry.

Therefore, by taking the manufacturing industry as a media, Apple Inc. can not only maximize the investment return, but also minimize the business risks and optimize the investment exit mechanism. Meanwhile, by transforming itself into a leading financial investment company, investment efficiency is optimized, investment success rate is maximized, and investment risk is minimized. Finally, Apple Inc. achieves the optimal integration and innovation of the economy, finance, industry, technology and service. No wonder the share of Apply Inc. is so popular in USA's capital and stock markets. The current market value of Apple Inc. has reached about 800 billion US dollars, making it the largest over the world.

Indeed, Apple's "light-asset" industrial development model is based on the standards, software, Internet, big data and AI. Such model has significant impacts on the industry and enterprises. The core value of big data can be now better understood. In such new industrial mode, the information technology (e.g. software, Internet and big data) is deeply integrated with all sectors. So, the first role of information technology as production tools is to improve the productivity. The efficiency of enterprises and industries is greatly improved whereas the costs are significantly reduced. Then, the information technology revolutionically transformed the production relations and industrial development model. It can be observed that the Apple Inc.'s business crosses a variety of industries such as: manufacturing, service, chip design, basic software, application software, industrial software, Industrial Internet, Internet, big data, AI and even the financial investment. This represents a revolution to the traditional industrial form and ecology. Similarly, the Internet, big data and AI economy emerged. This is also the future direction of Chinese enterprises, industries and economy.

Traditionally, all industries (e.g. manufacturing, service, software, Internet, telecommunication, financial, insurance) are separated in clear boundaries for the purpose of effective administration regulation. Today, the Internet and big data oriented new economy has historically transformed the production relations, production mode and economic structure, in which multiple industries will be deeply integrated. The nation's superstructure is also required to make a substantial adjustment and transformation accordingly. For the national economic base, it is necessary to provide a macro-environment which is based on the Internet, 5G, big data centers, AI and cloud services. Such macro-environment must provide high efficiency and low cost services to the entire society and industries.

The Apple Inc. takes the leading role in the new economy and the new industrial development model. The followers include Boeing and GE (USA), Siemens (Germany), the Rolls-Royce (UK) and a large number of Internet companies. As another follower, Beijing's MI is the only Chinese company which adopts the industrial Internet development model. There have been fundamental changes to the world economy. So far, a large number of the Chinese listed enterprises still have adopted the traditional industrial development model. These Chinese enterprises are feeling enormous pressure from the investment and capital sector, as the market value of their shares are declining continuously. Indeed, it should be understood that "Internet" is not a technical term anymore. Rather, it has been evolved into an economic concept and served the new methodology for the development of economy, industry and enterprise. Consequently, new production relations, new industrial development model and new economic form emerged. The economic, industrial and enterprise development has gradually and closely been integrated with the "Internet", software, big data and AI. After eight decades, the industry has fallen in a deep "love" with the software sector, Internet sector, big data and AI.

In fact, the Industry 4.0 era defined in this chapter is similar to the Internet economy era proposed by the third industrial revolution in USA.

Chapter 2
The "Chemical Reaction" Derived From the Interaction Between Internet and Economic Methodology

In economic activities, if the innovation and application of a core technology can lead to the transformation of the industrial development mode, its impact will be extremely far-reaching. The Internet has led to and perfectly realized the new mode of industrial development, i.e. "light-assets". This will result in significant change in the economy and industry.

In the first chapter, we discussed the relationship among computers, software, the Internet and industry in last 80 years. In particular, the deep integration of computers, software, Internet and manufacturing sector has profound impacts on the economic, social and industry development. In this chapter, from economic perspective, we will discuss the "Chemical Reaction" derived from the interaction between Internet and economic methodologies as well as its significant impacts on various sectors such as: manufacturing, service, financial investment, software, and Internet. As a consequence, the values of the economy and industry have undergone radical changes.

The core of economic activities lies in the investment, namely, the capital contribution. In essence, human beings' production activities, marketing activities and indirect investment activities are a kind of investment. In addition, the fundamental of investment activities is fulfill the expectations, e.g. satisfying the market demand. The bare minimum expectation is to avoid loss via innovation on technologies, industrial development mode, service quality and investment methods. The higher expectation is to achieve profit, i.e. a better return on investment. The highest expectation is to maximize the return on investment (i.e. profit). The history of human development for thousands of years and the history of industrial development for more than three centuries can be summarized into the close-loop of investment, fulfilling expectations and maximizing the return on investment through innovations. This is the most fundamental principle of economic theories.

In order to encourage investment, fulfill expectations, and maximize the return on investment, human beings are constantly searching for new technologies and methods at different stages of development, inventing numerous technologies and developing countless products, achieving innovations in a variety of management

J. Yuchi, *Value Proposition*,
https://doi.org/10.1007/978-981-15-5025-6_2

methods. Similarly, the productivity is continuously being improved while production relations are adjusted accordingly. These have constantly enriched the economic theories. This period witnessed the first industrial revolution and the second industrial revolution. The human development history and economic development theories have gained an unprecedented leap which indicated that a more civilized development stage of human race. In recent years, there have been a number of discussions on industrialization, informationization, development of high-end manufacturing industry and industrial development modes. This indicated that principles and theories of economic development have reached a higher level.

A conclusion can be drawn by summarizing the development process of economic theories. The best approach for the investors (enterprises) to achieve objectives such as: encouraging investment, fulfilling expectations, and maximizing investment benefits without a consideration of institutional, legal, policy and other factors, is that enterprises should be capable of continuously providing the market with innovation and design of the required products through technological and management innovation; maximizing the management efficiency; minimizing the cost; maximizing the responsiveness to the market demand; maximizing the customer satisfaction and service quality. As a consequence, the return on investment will be maximized and the associated risks will be minimized. However, this is easier said than done. To achieve this capability, on the one hand, enterprises need to invest heavily on land, plants, equipment and supporting facilities (i.e."hard" capability), and human resources, information system, corresponding supply chain and service links that are necessary for production activities and service activities. As a consequence, "heavy assets" are formed, which present potential risks to the investment. On the other hand, enterprises need to develop the capability to link all essential factors mentioned above together. As a result, the entire enterprise is adjustable, controllable and manageable (i.e."soft" capability). Therefore, enterprises need to locate a technology, a means, or a methodology to integrate all aforementioned aspects to form a unified platform for R&D, design, production, management, operation and service. For entrepreneurs and investors, it is in fact more difficult to develop the latter's "soft" capability than the former's "hard" capability. For a long period of time, it has become a dream of enterprises to achieve such integrated "soft" capability. In fact, a large number of entrepreneurs, capital investors and academics on economics and management have been struggling to pursue such technologies and methodologies that can systematically realize this dream.

Today, Industry 4.0 (originated from Germany) and Internet and Industrial Internet (originated from the United States) have become a heated topic in the industrial and capital circles globally. A careful study reveals that the technology and methodology that the vast number of entrepreneurs, capitalists, and economic management scholars have been pursuing have been reflected in the keyword of "Internet." It can be observed that the Internet is such a technology, means and a new methodology that can integrate all aspects of enterprises such as: resources, manufacturing processes, research and development, management, services, strategies, industrial technology systems, and industrial ecological chains, which means a new approach! More importantly, a very powerful economic tool is embedded in the Internet: through the deep application of standards, the Internet, software, big data, AI and innovations in management and service model, leading enterprises in the industry can largely outsource a large number of low & medium value-added activities (e.g. processing, manufacturing and service) to the downstream of supply chain and service chain. As a result, a new industrial ecosystem is formed through collaboration. At the same time, through the Internet and various types of software, the resources required for the production activities and service activities under such industrial ecology are deployed, controlled and refined. Consequently, through the analysis of the large data generated by the ecosystem network, the positive feedback is generated continuously. This will deeply influence enterprise's decision-making on strategy, industrial mode, plans and product development, design, techniques, manufacturing, processing, service, resource management, etc. As a result, the enterprise is gradually transformed from the traditional "asset-heavy" development mode (i.e. operates the products and enterprise itself) to the "asset-light" development mode (i.e. operates the core values of entire industrial ecological chain). Via such transformation, these enterprises will be located at the top of the industrial ecological chain, and operate all on the Internet platform, i.e. on the "cloud". Their goals turn to pursue the core technologies, core businesses, core values of the industry and the industrial ecosystem. They will be capable of achieving a multi-dimensional deep integration with the service industry, chip industry, software industry, Internet industry, big data-AI-consulting business and even financial investment industry. This is the so-called industrial Internet enterprise advocated by the United States during the third industrial revolution, which is oriented to some specific industries or sectors. In fact, this is the most fundamental component of the Internet economy in the United States. Via this approach, leading enterprises in the industry will transfer the risks associated with the traditional "asset-heavy" investment to those downstream enterprises in the industrial supply chain and services chain while distributing certain benefits. As a result, the investment and operational risks for leading enterprises can be shared within their industrial ecosystems and even the entire world. This greatly helps to avoid risks of "asset-heavy" investment derived from economic deceleration and downturn, thereby maximizing the interests of investors and fulfilling the expectations. "Asset-light", a novel industry development mode has emerged.

Nowadays, leading enterprises of the traditional manufacturing industry are transforming and upgrading themselves into "asset-light" industrial Internet enterprises, which is achieved with the deep support and application of the Internet, software, big data and AI. The "+ Internet" mode has emerged. By contrast, the Internet industry and software industry are constantly integrating and covering the traditional manufacturing sectors and associated sectors such as financial investment and service into the territory of real economy. The "Internet +" mode has emerged. These companies are also developing themselves into "asset-light" industrial Internet companies. In my opinion, the global industries, capital circles and economic circles have reached a consensus on these two new economic development modes. Therefore, it is widely recognized that efforts are required to support and develop the Internet industry and the "asset-light" industrial Internet industry. The implementation and promotion of these two new industrial development modes (i.e. "+ Internet" and "Internet +")

will eventually present a new complex industrial system or form under which any of leading enterprises may operate and cover various industries or sectors such as: manufacturing, service, chip, software, Internet, and financial investment on an unified platform based the Internet, big data and AI, i.e. the "cloud". With such a capability, these enterprises will have a core competency and will be able to turn themselves from a large-scale business to a powerful one. A nation will have capability of sustainable development and become an economic power if it has a large number of such enterprises or companies.

Apparently, an enterprise is transforming and elevating itself toward the high-end and the top of the industry value chain if it has turned to the "asset-light" industrial development mode. However, this doesn't mean that those enterprises following the traditional "asset-heavy" mode do not exist anymore. On the contrary, a large number of such enterprises are located in the middle and low-end of the industrial value chain. They provide necessary support to those leading enterprises of the industry that adopt "asset-light" development mode. As a consequence, an industrial ecology is formed.

For China, a typical country with major economic structure dominated by "asset-heavy" development mode such as medium- and low-end processing, manufacturing and assembling industry, the only way for leading enterprises' sustainable development is to transform and upgrade themselves as soon as possible through the adjustment of the industrial structure. These include: the integration, transformation and upgrading of traditional manufacturing, financial investment, and service industries; the support and application of the Internet, software, big data and AI; appropriate use of both new economic development modes (i.e. "+ Internet" and "Internet +"), and vigorously developing the "asset-light" industrial Internet industry. Meanwhile, the state and governments at all levels are required to re-examine and re-establish their understanding of the industrial structure and format, to adopt strong policies and measures through reform and innovative practices, to vigorously promote the development of the complex industrial format. As a result, China's economy can be in line with the general trend of international economic development, and with the new mode of Internet economy. China should become a practitioner, promoter and leader of the Internet economy.

It can be considered that the Internet has become a most powerful technology, tool and methodology for solving the key problems that have long plagued the manufacturing, financial investment, and service industries. Indeed, the Internet is not simply a technical concept. Today, the word "Internet" has become a new and key economic theory, or the latest and brightest component of economics. Due to the emergence of the new theory of the Internet economy, in essence, major reform in productivity have triggered a significant adjustment in the production relations. Similarly, the economic and industrial values have changed dramatically. In my opinion, this is a radical "chemical reaction" derived from the interaction between Internet and economic methodology, i.e. industrial revolution. Under which, the format of global economy, manufacturing industry, financial investment industry and service industry will undergo profound changes.

So, all problems and challenges currently associated with China's economy, industry, enterprises and stock markets reflect the ripple effects and results of this profound change. At the same time, due to such changes in production mode, production relations and industrial development modes, it is necessary to adjust and reform China's economic base and superstructure as quickly as possible so that they are compatible with the development of the new Internet economy and new industrial development mode. In other words, challenges and opportunities coexist.

Therefore, it is imperative to clarify the Internet, the industrial Internet, Industry 4.0 as well as the new industrial development mode as the consequence. This is essential to discuss the economy, financial capital, the development of manufacturing industry and stock market today.

Chapter 3
Interpretation of Germany's Industry 4.0 From the Perspective of Economy and Industry

To better understand Germany's Industry 4.0 and its rationale, it is necessary to discuss the logic and meaning of standardized, digitized, cyberized and intelligent manufacturing in the first instance. Then attention should be paid to the Internet economy, a new growth model featured with "de-industrialization" and "asset-light".

In recent years, Germany's Industry 4.0 has been a hot topic in China. There is lack of unified report for Industry 4.0. Rather, various descriptions have been made by a number of German experts from different perspectives, which are difficult to understand. As a result, the understanding of Industry 4.0 varies according to individuals and industries. Some people perceive that Industry 4.0 represents automation factories, unmanned factories and automation workshops. The opinion is reinforced by the fact that Siemens (a Germany company) has recently built a demonstration workshop in Chengdu, China. Some people perceive that Industry 4.0 represents robots and robot production lines. Some people think that Industry 4.0 represents a CNC machine tool network. Some people perceive that Industry 4.0 represents a digital manufacturing and networked manufacturing industry. However, these are not new concepts and technologies. There have been a large number of cases and practices for such workshops, factories and technologies in the United States, Germany and China. Is Industry 4.0 simply a summary or integration of all concepts and technologies that we have been familiar with?

Others believe that Industry 4.0 represents a more advanced intelligent manufacturing. Such understanding seems to be closer to the definition of German Industry 4.0. However, the question is, what is intelligence? What is smart manufacturing? Is it really such a simple question? Is Germany's Industry 4.0 related to the Internet economy launched by the United States?

J. Yuchi, *Value Proposition*,
https://doi.org/10.1007/978-981-15-5025-6_3

1 Germany's Industry 4.0 and New Industry Development Mode

In my opinion, there are three fundamentals to solve the mysteries of Germany's Industry 4.0.

First: standardization and digitization.
The future's manufacturing will definitely be a large-scale environment and system which integrates manufacturing equipment, facilities, network, software and data. In such environment, hardware and software, data and knowledge, management and service will be integrated through the network. Therefore, the main manufacturing equipment, facilities and instruments need to be digitized. In other words, every piece of hardware, facility, equipment, instrument needs to have a virtual digital representation and definition. It is comparatively easier to achieve partial digitization. However, it is not meaningful for the future. To achieve overall digitization of a complex and massive manufacturing environment is never an easy work, its precondition must be based on a set of uniform standards, specifications, and data dictionaries. Therefore, for a future interconnected manufacturing system, it is crucial that the manufacturing digitization is based on unified standards and their derived digital standards as well as consequent interconnection.

The first priority of Germany's Industry 4.0 is to develop standards and digital standards for all manufacturing industries in Germany. This is a fundamental task which involves a large volume of works. It requires a large amount of human and financial resources from the nation, the industry, and especially the leading enterprises in the industry. It will take several years to accomplish.

It is not accidental that Germans are so concerned about standards and digital standards. The standard is essential for a manufacturing industry or a leading manufacturing enterprise to achieve the digitized and cyberized manufacturing. More importantly, an enterprise can develop an industrial technology system with itself as core once it has mastered the industrial standard. Consequently, a leading manufacturing enterprise can make most of networks. Internally, it can achieve the interconnection of all four main processes, i.e. R&D, processing and manufacturing, management and service. Externally, the enterprise is capable of rapidly cooperating with global partners to build an industrial ecosystem. As a result, rapid expansion is achieved.

The standardization and digitization is a cost, a process and a foundation, rather than a purpose of Germany's Industry 4.0. Indeed, Industry 4.0 provides necessary foundation for the cyberized and intelligent manufacturing in the German industry, as well as the new growth model of Internet economy.

Second: cyberization and intelligent manufacturing.
With the foundation of standardization and digitization, leading manufacturing enterprises in Germany can use all types of software and operation management platforms developed by themselves to achieve the interconnection of their all internal hardware,

system, data and various resources and that of external industrial eco-chains, thus achieving the cyberized manufacturing. A large number of small and medium-sized enterprises will be integrated into such industrial networks, thus forming a new industrial ecosystem.

Let's have a close look of the manufacturing and management mode adopted by Airbus. A large number of enterprises from different countries are coordinated to design, process and manufacture millions of various parts and components for a complex aircraft. Airbus is in charge of the final assembly, sales and service. Such manufacturing and management mode is not possible without the support of standardized, digitized and cyberized production. It can be observed that Airbus is a leader in standardized, digitized and cyberized manufacturing.

In fact, Germany's Siemens is also doing the same work and making the same efforts. Despite with strong foundation of software, information and automation technologies, Siemens feels it is lacking in software and network technologies compared to counterparts in United States. Therefore, Siemens has acquired several software companies over the last decade. Table A.1 in Appendix A presents a summary of Siemens' acquisition of software companies all over the world. It can be observed from the table that the company has been continuously to focus on its informationization so as to realize the standardized, digitized, cyberized and big data-oriented manufacturing.

A large amount of data will be generated from industrial ecosystem based on the interconnection of networks. Can the data produced by a manufacturing enterprise itself be called the big data? The answer is "no". Only the complete valid data generated by an industrial ecosystem which is constructed by a network-based and industry-oriented leading manufacturing monopoly enterprise can be described as the big data. In other words, big data has attributes of the specific industry.

With big data in place, data can be categorized through statistical methods for processing, mining and analysis. As a result, data will be transformed into the knowledge, and consequently a knowledge base will be established. Previously unknown disciplines may be identified while such new knowledge will be shared across the network (i.e. Knowledge Explosion). With such a new method based on the big data, we will have new cognition and understanding of all manufacturing processes such as the R&D, technique, processing, management and services. The knowledge-based manufacturing will lead us to our desired intelligent manufacturing and smart manufacturing.

There is a clear logic: digitization will be achieved based on standardization; cyberized manufacturing will be achieved based on digitization; big data will be achieved based on cyberized and interconnection of networks; knowledge will be acquired based on big data; smart manufacturing will be achieved based on knowledge. In short, the Germany manufacturing industry is heading towards the direction of the standardized, digitized, cyberized, big data oriented and intelligent manufacturing. By 2030, a new format and new capacity of the Germany manufacturing industry as well as its industrial ecosystems will be displayed to the world.

Perhaps the traditional manufacturing industry will be satisfied once the so-called intelligent manufacturing is achieved. However, that only means the half journey for the Germany's Industry 4.0. From the perspective of catching up the Internet new economic theory and new industrial development mode, I believe that a solid foundation is just made for the transformation and upgrading of the manufacturing industry.

Third: "de-industrialization" and "asset-light", the Internet will result in a new industrial development mode and a new economy.

Once a leading manufacturer possesses these "soft" foundations and support (e.g. the standardization, digitization, networks, big data and intelligent system), those long-standing bottlenecks restricting the growth of manufacturing enterprises (e.g. the open source, open collaborative design, highly flattening management, customized scale production, personalized services, product life-cycle service, building of industrial supply chain and ecological chain) will be easily solved. A revisit of situation revealed that such a manufacturing enterprise has gradually transformed from a company that is solely engaged in production and manufacturing to an industry leader that relies mainly on networks, software, and data, as well as managing and controlling everything of the enterprise itself and the self-centered industrial ecosystem. With such a capability, this enterprise will inevitably outsource the processing, manufacturing and service with medium and low value through its industrial ecological chain, i.e. "de-industrialization". Meanwhile, such manufacturing enterprise will inevitably move up to the high end of the industrial value chain, gradually transforming and upgrading itself to a so-called industry-oriented Internet enterprise or industrial Internet enterprise. The development of the enterprise will be shifted from the traditional "asset-heavy" investment growth model to new "asset-light" growth model that is obviously supported by the Internet, software, big data and AI. The fundamental of "asset-light" is to control the industry's core technologies and core businesses. Such model is oriented to consulting, finance, management and service where the target is achieving core values of the industrial ecosystem. As a result, the return on investment can be maximized while the associated risks can be minimized. Such transformation from a manufacturing enterprise to an industrial Internet enterprise can be called the "+ Internet" model today.

It can be expected that Germany's Industry 4.0 will adopt the "+ Internet" model by taking the advantages of the Germany manufacturing industry's superiority in the world. As a result, the transformation from the traditional economic development mode to the new development mode of Internet economy will be achieved.

2 How Far-Reaching Impact the Industry 4.0 Could Produce on the German Economy and Enterprises?

For a long period of time, the leading manufacturers have been responsible for the integration and allocation of resources within the industry. For instance, leading manufacturing enterprises are in charge of the R&D, production, assembly, sales and service of products. However, such a traditional manufacturing growth mode will be gradually replaced by a new growth mode of the Internet economy which outperforms in efficiency, cost, agility and service.

In the new development mode of Internet economy, the integration and allocation of resources within the manufacturing industry, as well as the production, assembly, sales and service of products will be gradually controlled and managed by the Internet or industrial Internet companies. In this mode, the manufacturing enterprises and products will become more specialized. The classification of disciplines will become increasingly elaborated. Meanwhile, the manufacturing resources and activities will be more and more modularized in the Internet business mode. The manufacturing activities will remain the same, but the traditional manufacturing industry will gradually become processing industries whose profit margins will be successively taken by the industrial Internet businesses. This will surely lead to the financing difficulties and decline of share price for traditional manufacturing enterprises. These traditional manufacturing enterprises will eventually become a portion of the Internet industry's chain and become increasingly reliant on Internet and industrial Internet companies. In this mode, the industrial Internet companies or Internet businesses will gain widespread popularity in the capital circle and easily attract substantial amount of capital investment. Meanwhile, their profit margins and share prices will increase continuously. Indeed, these companies will gradually dominate the manufacturing industry.

In new development mode of Internet economy, instead of making direct investment to manufacturers, the finance sector is more inclined to make indirect investment through the industrial Internet companies or Internet businesses. Therefore, the Internet economy will not only change the traditional operation mode of the manufacturing industry, but also transform the traditional operation mode of finance sector. This will result in substantial changes to the manufacturing industry and finance sector, which will lead to an industrial and economic revolution.

Germany has witnessed the following facts: the continuous advancement of a multitude of Internet and software companies from United States into the manufacturing industry, including Apple, Google, Microsoft and so on. These companies have extended their business to various sectors of the real economy such as automobile, digital home appliance, digital health care, robotics and energy production, which are traditionally the territories of manufacturing industry. By continuous extending businesses into the Internet industry and the "cloud", these companies are transforming themselves into the industrial Internet companies. Meanwhile, traditional manufacturing enterprises of the United States (e.g. Boeing and GE) are increasingly growing their business by capitalizing on the software, Internet, big data and AI. In addition, the Chinese manufacturing industry has gained a rapid development, taking a growing share of the global manufacturing market. Facing with such great industrial and technological changes, Germany has great concern that its predominant manufacturing industry will be taken over by those powerful Internet companies and industrial Internet companies from the United States, and consequently become a part of their supply chains and ecosystems. Under enormous pressure, Germany has determined to take the initiative to transform the manufacturing industry through the implementation of Industry 4.0 strategy. Since then, Germany's manufacturing industry has commenced the process of informationization and Internet orientation by means of adopting the "+ Internet" mode.

A thorough analysis of Siemens, the most typical enterprise in Germany revealed that it remains a monolithic "asset-heavy" enterprise. Siemens has outsourced a large amount of low value-added activities such as processing, manufacturing and services in global scale. However, this enterprise still hasn't completed the transition from the mode of running its own business to the mode of operating core values of the industrial ecosystem. In fact, Siemens hasn't completed the transformation from the "asset-heavy" oriented development mode to the "asset-light" development mode. This indicated that Siemens hasn't accomplished the comprehensive transformation to an industrial Internet company. Therefore, I believe that Siemens will continue to substantially "lose weight", by downsizing and restructuring towards an "asset-light" oriented industrial Internet company in the near future.

According to the overall economic quality of Germany, vast majority of enterprises have completed the industrialization process. In addition, a large number of German enterprises have occupied the high-end manufacturing industry, outperforming manufacturers from other countries in the global market and controlling the profit margins. These are the superiority of Germany and its enterprises. Their short slab is that there are a fraction of enterprises which haven't completed the informationization process. Indeed, the biggest problem is that Germany's global-leading manufacturing enterprises are dominated by businesses predominately investing in "heavy" assets. These enterprises have not achieved the transition to the "asset-light" mode. Although the Germany manufacturing industry has an obvious cutting edge globally at the moment, the overall national economy is centered on heavy investment and is likely to encounter problems in the future. Therefore, the heavy investment and "asset-heavy" development mode in the manufacturing industry have presented significant challenges to the economic structure of Germany, where urgently needs solutions and mitigation strategies. In fact, the right solution and the correct choice is that Germany enterprises adopt "+ Internet" mode and transform themselves into "asset-light" industrial Internet companies through the standardization, digitization, network orientation and intelligence orientation as soon as possible.

Another critical structural issue associated with the Germany economy is that its software industry, especially the Internet industry, is lagging far behind their counterparts in United States. Similarly, Germany's Internet industry lags far behind their counterparts in China. This is clearly demonstrated via an analysis of the world's top 10 software companies, the world's 20 highest market-value Internet firms and the world's top 25 IT enterprises (see Tables A.2, A.3 and A.4 in Appendix A). As shown in three tables, United States is in the monopoly position in global IT industry, software industry and Internet industry. Only second to the United States, China has a total of 6 firms that are listed in the world's 20 highest market-value Internet firms, while Germany has zero. Very small number of Germany companies are listed in the global top 10 software firms and top 25 IT firms. Failing to close this gap in a short period of time, it is likely that Germany will lag further behind the United States in Internet industry and industrial Internet industry. Consequently, Germany's strong manufacturing industry may be merged into the ecosystems of the Internet industry and industrial Internet industry of United States, becoming its sub-sector with main focus on processing.

Therefore, in essence, there are three purposes for Germany to implement Industry 4.0. The first purpose is to complete German enterprises' informationization process as soon as possible to deal with its short slab in "software". The second purpose is to complete the transformation and upgrading of the Germany manufacturing industry from the traditional manufacturing mode to the industrial Internet industry mode as soon as possible. As a consequence, the German economy and industries can achieve the transition from the traditional manufacturing industry's "asset-heavy" development mode to "asset-light" oriented industrial Internet development mode. This will help the Germany manufacturing industry to control the global industry ecosystem, to occupy the high end of manufacturing value chain, to maximize the return on investment and to minimize the investment risks. The third purpose is to improve the overall quality of national economic operation, to maintain the leadership and dominance of the nation and its enterprises in the world, and to deal with challenges presented by the changes derived from the Internet economy of United States. Meanwhile, through the formulating and implementing the United Nations Convention on Atmospheric Emission Reduction, Germany can occupy the moral high ground in the industry and maintain the advantage position. Together with the United States and other western powers, Germany can control the development speed of the emerging and developing economies (e.g. China and India) and their enterprises, thus continuing to maintain its economic power and hegemonic position.

In essence, there is no fundamental difference between Germany's Industry 4.0 and United State's Industrial Internet.

It can be predicted that, enterprises' production activities and human being's social life in the future are manifested to be associated with a wide variety of Internet service providers or through the industrial Internet companies in varies industries. Smart phones, smart terminals, 4G, 5G and all types of App software services & AI can be seen everywhere. However, in reality, the manufacturing industry will be gradually taken over and integrated by Internet businesses and industrial Internet companies.

Therefore, manufacturing was the dominant industry in the past, which however will be replaced by the industrial Internet industry and Internet industry in the future.

It is understandable that Germany proposed the concept of Industry 4.0, rather than the Industrial Internet. The concept of Industrial Internet concept was proposed by the United State. As one of the world leaders in manufacturing, the morale of the Germany manufacturing industry and IT industry can be boosted with its own concept, i.e. Industry 4.0. Via this approach, Germany can avoid consolidating opponent's power and prestige. However, in essence, Industry 4.0 represents the same thing with the Industrial Internet.

According to the above analysis, Industry 4.0 is industry-specific. In other words, Industry 4.0 is a major task of the manufacturing industry and global leading manufacturers in Germany. Once the Industry 4.0 is completed, small to medium sized enterprises(SMEs) in Germany will become integral parts of the different industrial ecosystems according to industrial standards, software and the "cloud" developed and established by those German leading enterprises. This is also the reason the vast majority of German SMEs do not have a clear understanding of Industry 4.0.

Chapter 4
The Significance of the American Third Industrial Revolution Based on Internet and Big Data

Why was the United States so determined to promote the third industrial revolution based on the Internet and big data? What are the essence and fundamentals of the third industrial revolution? What forms would the new industrial development mode of the Internet economy present? What are profound impacts of the third industrial revolution on the traditional manufacturing industry, financial investment industry, service industry and the stock market?

The software, Internet, big data, AI and other information technologies have result in the industrial transformation. This has led to significant changes to the mode of economic growth as well as the value system of economy and industry. This development has been unfolding in the world. Therefore, it is necessary to discuss the revolution derived from the Internet, industrial Internet and big data, namely, the third industrial revolution.

1 Those Who Conform to the Industrial Revolution Will Achieve Success

In general, the economic development is mainly driven by the market. However, each industrial revolution was driven by a significant technological innovation. The emergence of new productive forces led to the innovations and changes in the production mode and industrial development model, the adjustment in relations of production, and changes and advances to the social systems and lifestyles.

The first industrial revolution was driven by the invention of steam engine technologies. To simplify, the invention of the steam engine technology led to the emergence of electricity, which is the energy necessary for the mechanical equipment. Due to the easier accessibility to electricity, a large number of machinery manufacturing enterprises were established and sustained rapid development. From the perspective of economic and social form, the invention of steam engine technology led to the transformation in economic development mode and the collapse of feudal society.

J. Yuchi, *Value Proposition*,
https://doi.org/10.1007/978-981-15-5025-6_4

Those landlords did not play a critical role in society, production, marketing and investment activities anymore. By contrast, entrepreneurs and capitalist emerged. Steam engine and electricity generation technologies were invented, re-invented and innovated by the Europeans. Their society was the first advancing from the feudal social system into the capitalist social system, leading the rest of world by more than three centuries.

The second industrial revolution was driven by the electrification technologies. The emergence of electrification technologies allowed manufacturing enterprises to have rapid mass production of the goods needed by the society, the raw materials and products needed by other enterprises within the supply chain. As a result, the society's needs were greatly satisfied as the commodity was supplied in abundance with low prices. From the perspective of economic form, a large number of small to medium-sized workshop-style manufacturing enterprises and service enterprises were replaced by a pack of large-sized transnational manufacturing companies and services giants. Similarly, an increasingly number of industry leaders and oligarchs emerged. As a result, the economy, industries and enterprises achieved a giant leap from the localization to the globalization, which has reshaped the global economy. From the perspective of social form, the emergence of the electrification technologies provided the solid support to the prosperity and stability of the Western capitalist social system. Meanwhile, the status of the Western powers was increasingly consolidated.

Today, we are experiencing a transformation that is led by Apple, Google, Microsoft and other American Internet and software companies, and is driven by the software, Internet, cloud-computing, big data, AI and other information technologies. Can this transformation be called as the third industrial revolution? The answer is: Yes. The emergency of information technologies, particularly the Internet, industrial Internet, big data and AI, will result in earth-shaking changes to the society, economy, production and life.

2 What Form Would the New Industrial Development Mode of the Internet Economy Present?

In fact, the third industrial revolution has been kicked off since the early 1990s when the Western countries led by the United States proposed the economic globalization. In the economic globalization, the leading manufacturers in the United States and other Western developed countries started to outsource the medium and low-end activities such as manufacturing, processing, and services. Their manufacturing and service capabilities may seem being weakened. In reality, they are occupying the high-end of the industry chain. With a great control of the global industries, their profits are super high. It is not straightforward for those western leading enterprises to achieve such the outsourcing of manufacturing, processing and service at the global scale. This is due to a large number of challenges in terms of technology, management, and implementation. Indeed, such process requires the strong support of information technologies, such as standards, software, networks, cloud-computing, big data, AI, etc.

In recent years, driven by the global expansion and competition needs, these transnational enterprises in United States and other Western developed countries are increasingly reliant on information technologies. With strong supported of the information technologies such as software, Internet, big data and AI, these enterprises are constantly advancing to the high-end of industrial value chains. Meanwhile, these enterprises are constantly strengthening the construction of the self-centered industry technological systems and associated industrial ecosystem. Through building their own industrial Internet form, these enterprises are aiming at maintaining their monopoly and competitive advantages.

In the form of industrial Internet, the leading enterprise will focus on the construction of self-centered industrial technology system and industrial ecosystem. This can be achieved through the development and control of digital standards, software, network, large data and AI. Their focuses are placed on: the deep fusion between the industry and the finance; the control of the industry's core technologies and the protection of intellectual properties; the allocation of global superior resources; the product's innovation and design; and the high value-added work such as manufacturing and service of high-end products. On the contrary, a large amount of medium and low value-added manufacturing and service activities will be undertaken through its industrial supply chain and industrial ecosystem. Therefore, in the future, a leading enterprise will seem more like a multi-dimensional "asset-light" complex specialized in a certain industry or sector. It is a complex of industrial standard enterprise, high-end manufacturing enterprise, software company, big data business, Internet company, financial consulting business, service enterprise, and management consultancy. To put it simply, such an enterprise can be called as an industrial Internet company with an aim at pursuing core values of that industry. Such a company would occupy the high-end of the value chain. Meanwhile, by controlling the industrial technology system and the industrial ecological chain, global superior resources will be allocated for collaborative design, manufacturing, management and service.

Thus, the industrial Internet will have a significant impact on the sectors, industries and economy. Different from the "asset-heavy" industrial growth mode adopted by traditional manufacturing enterprises, the growth of such industrial Internet companies is based on "asset-light" industrial development mode supported by the Internet, software, big data, AI and other information technologies. In this mode, the industrial Internet company pursues the industry's core business and core values. As a result, the efficiency, profits and return on investment will be maximized; the management costs, the response time to the market and the investment risks will be minimized; its services will be optimized. Its dominating position will be maintained in the intense market competition. In their original "asset-heavy" manufacturing mode, companies in the United States and other Western developed countries produced the huge volume of pollutant emissions. In such new "asset-light" development mode, these western companies will consequentially transfer massive pollutant emissions, together with a great deal of medium and low-value activities on processing, manufacturing, assembly and services, to the emerging and developing countries. As a consequence, they can occupy the moral high ground in economic development in a bid to continue controlling and constraining the economic structure and growth speed of developing countries. Thus, they can achieve the target of continuing to lead the development of global economy.

The economic, capital, and industry circles in United States and other western developed countries have reached the common understanding that the "asset-light" industrial Internet businesses should be supported. In the United States, an increasingly number of traditional leading manufacturing enterprises have consciously transformed into the industrial Internet companies. They play a major role in the capital markets and stock markets of United States. Such enterprises are more likely to receive heavy investment from the capital circles. As a result, their shares are enjoying a high popularity. In Europe, following Germany's Industry 4.0, a growing number of leading manufacturing enterprises are striving to achieve this objective.

In fact, the competition for the industrial Internet comes from two different camps, i.e. the manufacturing industry and the Internet industry.

Based on its own strengths in manufacturing resources and technologies, the manufacturing is marching towards the industrial Internet industry, i.e. "+ Internet" model. They are the main driving force of the national economy. Apart from the United States, enterprises in most countries need to adopt the "+ Internet" model. The manufacturing industry is a major pillar of the national economy in China. Therefore, "+ Internet" should be the main economic development mode of China in the future.

Based on their own strengths in information technologies and financing, the software and Internet industries are marching towards the real economy and the industrial Internet, i.e. "Internet +" model. They are new driving forces of the economy. An analysis of Table A.2 (Global Top Ten Software Manufacturers) and Table A.3 (The World's Top 20 Internet Firms with Highest Market Value) in Appendix A clearly shows that a large number of companies are from Unites States, such as Google, Apple, Amazon and Microsoft. These companies are dominating the global technologies and industries in software and Internet. As predominant and invincible leaders in the "Internet +" model, these companies are making significant impacts on the traditional service and manufacturing industries at the global scale. The US Internet companies' advantageous and monopoly positions in the "Internet +" model has drawn concerns from not only the US traditional service and manufacturing industries, but also the world's major economic powers and their enterprises. Germany is now making responses by actively implementing the Industry 4.0 strategy. Similarly, some Chinese Internet companies such as Alibaba, Tencent, MI, Baidu, 360 and Didi have made considerable achievements through continuous innovation of their technologies, services and operation model.

At present, the competition between "+ Internet" and "Internet +" is increasingly intensified.

Representatives of manufacturing enterprises adopting the "+ Internet" model include: Boeing and GE in the United States, Siemens in Germany, Airbus in Europe, and Rolls-Royce in the United Kingdom. In the past two decades, Boeing's development has become increasingly reliant on the software and informationization. Fully digital drawings and information management was achieved in Boeing 777 while all traditional manual drawings were eliminated. Boeing has fully integrated the R&D, manufacturing, management and services. Following the unified standard, its design, manufacturing, management and services of Boeing 787 were completed through the collaborations of various enterprises that are from 48 countries. Both operation management and services were undertaken through the network where more focus is placed on the customer's experience. Boeing is a leader in transforming itself from a manufacturing enterprise into an industrial Internet company, namely the "+ Internet" model. Its philosophies of design, management, manufacturing and service are most advanced in the world. As a world's leading provider of high-end engine products, GE has undertaken a software-based and informationized development process in the past two decades. The processing and manufacturing of components of GE's high-performance engines are increasingly outsourced. Even the core thermal components are completed through collaborative design and manufacturing. Indeed, GE has tried not to be involved in the hardware processing and manufacturing. Various software was developed and used to solidify the R&D of core technologies, engine products design, technique and manufacturing, engine maintenance and repair services, enterprise resource management, data and knowledge base and many other links. Its resources allocation and management are providing the strong support to the synergistic operation of the supply chain, service chain and ecological chain. Today, GE is more willing to declare itself as a software company. It seems that GE is a typical industrial Internet company in the field of high-performance engines and a leader of the "+ Internet" mode.

Following the "Internet +" mode, the Internet and software companies are continuously moving toward the downstream of the supply chain and marching into real economic sectors. These companies are rapidly entering the territory and market of traditional manufacturing, and then advancing into associated service industries and towards industrial Internet industries. Typical examples of these companies include: Google, Apple, Microsoft, and other US Internet and software companies. These companies are constantly advancing into the automobile, robot, digital home appliances, energy and medical health industries and so on. Similarly, they are continuously integrating the manufacturing industry with associated service industries. Due to the impacts of Internet companies, the form of traditional manufacturing industry and its associated service industry that we used to be familiar with are undergoing the revolutionary changes. This indicates the emergence of the Internet Economy Age that is represented by the "asset-light" industrial Internet industry.

The two camps, the manufacturing industry and the Internet industry, are marching towards each other. In the competition towards the industrial Internet, in general, the "+ Internet" model of manufacturing industry is on defensive, while the "Internet +" model of the Internet industry is on offensive. Such findings could be easily explained by studying the US stock market. The share prices of the manufacturing businesses are 6 dollars per share on average and that of a performing software company could reach 60 dollars. However, a share of an Internet company such as Google or Apple could cost as high as $600. In the first day of being listed in the US stock market, China's Alibaba's share price reached about $100 per share, financing 25 billion US dollars. The share prices in Internet companies are two orders of magnitude higher than that of manufacturing enterprises. With abundant cash in pockets, those Internet companies are fully confident and have endless opportunities to integrate manufacturing and service resources. On the contrary, the traditional manufacturing businesses are increasingly losing favor of the financial capital sector. They are under enormous pressure for the survival and development. Therefore, it seems to be inevitable to transform to the "asset-light" industrial Internet industry as soon as possible if the manufacturing industry wants to improve its fortune for survival and development in the future and not to be controlled by the Internet industry.

It is worth noting that, it is a restructuring, upgrade and transformation process and an inevitable development trend for the manufacturing industry to adopt the "+ Internet" model and the industrial Internet. However, it is a subversion and revolution to the traditional development mode of manufacturing and service industries when the Internet and software industries adopt the "Internet +" mode and March into the real economy and the industrial Internet.

3 What Are the Fundamentals of the American Third Industrial Revolution?

As aforementioned, once the Internet companies (represented by Google and Apple) confirm the profitability of certain manufacturing sectors, they will continuously move to such sectors or industries. Striding into the real economy, they will penetrate into the manufacturing industry and then the associated service industry and financial investment industry. This indicate that these companies will finally March into the industrial Internet.

In this process, those Internet companies will rapidly integrate the global first-class manufacturing resources and industrial eco-chain. This can be achieved because of those Internet companies' advantages on superior financing capacity, super-high share values, abundant capital and the availability of digital standards, network, software, big data and AI. In contrast to the "asset-heavy" growth mode adopted by the traditional manufacturing businesses, the Internet companies will adopt the "asset-light" development mode to make most of the existing superior manufacturing resources and industrial eco-chain in the world. As a result, the Internet industry's management costs and their products prices can be greatly reduced, the efficiency and service can be enormously improved, return on investment can be maximized, investment risks can be minimized, and the competitive advantage can be maintained. If things go on like this, the traditional form of manufacturing will be completely changed where the existing manufacturing industry will be subverted. Therefore, the first fundamental of the third industrial revolution is that the Internet industry adopts the new economic growth mode of "Internet +" and "asset-light" to penetrate into the manufacturing and associated service industries, and to advance to the industrial Internet.

According to the Wall Street, the greatest concern of the traditional American banks is the challenges presented by the Internet financial service industry represented by Google, Apple and Amazon. In China, the traditional banks are most concerned with the challenges presented by the Internet financial service industry such as Alibaba and WeChat Payment. This is because these new Internet financial service companies have a mass customer base dominated by the younger that are the mainstay of future financial services. Today, a large number of outstanding talents are attracted to the Internet financial service companies from the traditional banks. This trend is inevitable. In other words, the Internet companies are moving upward (overall financialization) while the traditional banks will be moving downward. The only way of survival for traditional banks in the future lies in specialization and the close cooperation with the Internet companies and industrial Internet companies.

In the future, the Internet companies and industrial Internet companies will experience overall financialization, i.e. becoming the new banks or financial service institutions. By contrast, traditional banks will undergo revolution, being subverted by Internet companies. In other words, it can be expected that the Internet companies, industrial Internet companies and traditional banks will and should be merged, showing their strong financial attributes. This is another fundamental of the third industrial revolution. This also offers an explanation of the high share prices of Internet businesses and industrial Internet companies in US stock market (hundreds of dollars per share) from another perspective.

The development of industrial Internet might be seen as the intense competition between two camps, i.e. the manufacturing industry and the Internet industry. A thorough analysis shows that it is in fact another game of war and economy initiated by the United States and other Western economic powers against emerging countries and developing countries. The United States and other Western economic powers expect to take the leading position of global emission reduction through fighting for and leading the "asset-light" industrial Internet new mode. As a result, they can remain powerful and maintain their advantageous positions against developing countries. The aim is to control the pace and quality of economic growth of those developing countries and emerging nations. Through this war and game, the United States and other Western economic powers expect that their industries and services remain at the high-end of the value chain. Nevertheless, the developing countries are positioned at the middle and low-end of global industries and services. The United States and other Western economic powers attempt to let developing countries and emerging nations to continue engaging in the "asset-heavy" growth mode, i.e. traditional processing, manufacturing, assembling and low-value services, which is on the cost of valuable natural resources, continued pollutant emissions and environmental degradation. This is the fundamental and essence of the third industrial revolution.

From the perspective of economic form, the traditional manufacturing, service and financial industries were relatively independent in the past. Under the new development mode of the Internet economy, the Internet industry and industrial Internet industry will play the central role in the economy and be deeply integrated with the manufacturing, service and financial industries. The boundary between these industries will become increasingly blurred, forming a new complex industrial form. The Internet and industrial Internet will become a new key to open the door to future economy.

Google, Apple, Microsoft, Intel and other US companies maintain the monopoly and dominant positions in information technologies such as the Internet, software, cloud-computing, big data, AI and core chips. As a result, the United States leads not only technologies, but the transformations of economic development mode as well. This will result in profound changes to the form and pattern of the world economy. This change will be sweeping to the traditional manufacturing, financial investment and services industries, and even information-technology industry itself. Meanwhile, the industrial, economic and financial sectors and the stock market will be deeply affected. In such a transformation process, each nation, industry and business will not be able to avoid influences unless making adjustment accordingly.

It appears that the information technology (represented by the Internet, software, cloud-computing, big data and AI) is a technology leading the overall economy and industries rather than simply an ordinary technology. It is even a novel economic theory.

4 Which Industries Shall Be Transformed by the Unstoppable Industrial Revolution?

So, which industries shall be transformed by the revolution derived from the Internet, cloud-computing, big data and AI? The answer is that it will be a full-range revolution to all sectors relating the traditional manufacturing, service, financial investment industry, military and so on.

The traditional financial industry will be transformed. The United States is powerful in various aspects such as political, economic and military. Fundamentally these are derived from the financial strength of United States. The United States, a powerful financial empire, has experienced two strikes in the recent decade. The first one is the 2008 financial crisis. The US traditional financial industry is still struggling to recover from the crisis in the short term. This is due to the extremely imbalanced financial bond leverage and the lack of subsequent effective financing instruments. Meanwhile, there is a need to repay a large amount of overdue bonds and debts. The second one is the powerful challenge from and impact of the Internet industry. The only solution for the US traditional financial industry is to move "downward" by collaborating and to deeply integrating with the Internet businesses and industrial Internet companies. There is universal agreement by New York's Wall Street that the most competitive banks in the future will be the Internet industry banking represented by Google, Apple, Amazon. The Internet banking industry's charm and future lie in its massive number of young clients. They are users of Internet service today and will become loyal customers of Internet business banks tomorrow. Therefore, the Internet finance industry will completely rock the dominance of traditional US banks and investment companies, leading to a revolution to the traditional financial industry.

The traditional manufacturing industry will be transformed. With high share prices, the Internet companies possess super strong financing capacity. Internet companies have huge amount of capital and extreme familiarity with the information technologies such as software, Internet, big data and AI. As a result, Internet companies can achieve the rapid integration of global manufacturing resources to penetrate into the manufacturing industries through their "asset-light" industrial Internet development mode. Future of product design will be conducted through the Internet to achieve the maximum collaborative design, open source design and customer's self-design. Design resources will be shared and greatly enriched. In the future, a highly flat management will be applied to the manufacturing resource management through the network to the largest extent. Similarly, point-to-point management and end-to-end management will be adopted in more cases in a bid to eliminate a large number of intermediate links and levels. As a consequence, the management efficiency will be significantly improved, and the management cost will be greatly reduced. With the support of big data and AI, the management will be more transparent, and decision-making will be more accurate and efficient. This helps to further reduce the management costs. In the future, manufacturing plants and manufacturing workshops will become increasingly specialized. Meanwhile, socialized division of labor will be increasingly meticulous. As a result, all manufacturing equipment, manpower and other resources will become a node in the network. The manufacturing process will be cloud-based while cooperative manufacturing will be based on Internet scheduling. Indeed, the customized manufacturing based on customer demands can be achieved. In the future, product services will cover the entire life cycle of products that are provided to the customer. The network, data, App software, AI and services will be ubiquitous. Therefore, the Internet and industrial Internet industry will transform the entire traditional manufacturing industries.

The traditional service industries have been or are being transformed. It is very convenient now to shop via e-businesses and online retailers such as Amazon, Alibaba (China) and JD.com (China). While chatting on Tencent's WeChat, you can use WeChat to shop according to your friend's recommendation. Apple and Tencent (China) can provide entertainment services. Baidu (China) and Google can provide answers to your questions. Uber and Didi (China) can provide taxi services. Internet can provide the services whatever you need. The Internet companies will provide everything involved in the Smart City and Smart Earth initiatives. Traditional service enterprises and service modes are gradually falling apart due to the impacts of Internet companies and their services.

The traditional military theories will be transformed. In the past, the military equipment was the leading factor. In today's Internet mode, the software, Internet, 5G, VR, AR, big data and AI are leading factors whereas military equipment become a foundation. The ground-air-space integrated war commanding and decision-making system based on software, Internet, 5G, VR, AR, big data and AI will determine the outcome of the war. This is what the US military is doing today.

Energy, transportation, healthy-care industries and so on will also be fully transformed.

It can be predicted that the foundation laid by the global top 500 enterprises in the past century will experience unprecedentedly enormous changes in the next decade. This is evidenced by the success stories of Google, Apple, Amazon and other Internet companies from the United States. The Internet companies from China (e.g. Alibaba, Tencent, Baidu, MI) will create a miracle by using the shortest time to become one of the top 500 enterprises with highest market value in the world.

Let's have a look at the business income of Fortune Global 500 firms in 2000 and 2015 (see Tables A.5 and A.6, Appendix A) and compare them carefully with Table A.3 of Appendix A. The most shocking fact is, in contrast to the Fortune Global 500 firms' century-long business history, those Internet companies (Google, Facebook, Alibaba, Amazon, Tencent and Baidu) only have a business history of 10 years, or 20 years. However, their capacity of accumulating capital and wealth is very impressive, so are their development speeds. The Internet companies will squeeze these traditional leading enterprises with a centenary of business history out of the Fortune Global 500 list and take over their businesses. The only solution for these traditional leading enterprises is to transform or upgrade themselves as quickly as possible following the orientation of Industry 4.0 (Germany) or Industrial Internet (USA).

Chapter 5
What Are Major Driving Forces to the US Manufacturing, Financial and Internet Industries?

What kind of development processes the US manufacturing and financial investment industries have experienced? Why are they so powerful? How did they locate the key for the "asset-light" industrial Internet growth mode and for the development of the Internet economy?

1 How Was the US Manufacturing Industry Created?

A detailed study of US economy shows that the US manufacturing has a history of more than a century which laid a solid foundation. This development process (i.e. the industrialization process) witnessed the continuous maturing of manufacturing technologies and intensifying competition. As a result, the socialized division of labor has become increasingly explicit and the manufacturing enterprises have become increasingly specialized. In such an intense competition process, a fraction of manufacturing enterprises have continue to restructuring, upgrading, elevating themselves to the high-end of the value chain of the manufacturing industry. These enterprises eventually become the leaders of the high-end manufacturing, e.g. Boeing, GE and other companies in the United States. Meanwhile, some manufacturing enterprises have transformed themselves into service providers of that industry while others have successfully transformed themselves into the top end of the industrial value chain, i.e. financial investors. In other words, a certain proportion of banks and investment companies in the United States are originally from the manufacturing sector. They have maintained a natural connection with manufacturing sectors. Therefore, their loan and investment businesses tend to be specialized and industry-specific. A large number of banks and investment companies in the United States are engaged in financial business yet drawing profits from the manufacturing sector. In fact, such a transformation represents another form, i.e. more advanced form of manufacturing industry.

J. Yuchi, *Value Proposition*,
https://doi.org/10.1007/978-981-15-5025-6_5

According to the above analysis, the United States and other Western developed countries have almost completed the process of industrialization along with the development of manufacturing industry.

Since the 1980s, driven by the powerful information technologies represented by US giants such as Microsoft and Intel, a large number of industrial software vendors and products have emerged that are oriented to the R&D, management, manufacturing and service of manufacturing enterprises. This indicates that the manufacturing industry in the United States and other western developed countries have initiated the process of informationization. This process has provided strong supports to the economic globalization, and the globalization process of large manufacturing enterprises has commenced. Subsequently, through establishing the industrial technology system and industrial ecosystem with the support of network and software, leading manufacturing enterprises in the US and other western developed counties started to outsource their medium and low value-added activities (i.e. processing, manufacturing and service) in order to constantly elevate themselves to the high-end of industrial value chain. In addition, through controlling core technologies and industrial standards, these enterprises implemented the integration of global resources to achieve maximum profits. In this process, GE and Rolls Royce adopted their business model of transforming from manufacturing to services. Boeing and Airbus adopted a model of collaborative design, processing, manufacturing, management and service at the global scale. Ford and Toyota adopted a global collaboration model that is featured with zero-inventory. Japanese and South Korean ship-building enterprises adopted their production model of reducing dock occupancy, segmental collaborative manufacturing and modular assembly. Through more than two decades of efforts, the major manufacturing enterprises of the United States and other western developed countries have completed or are close to the completion of the informationization process.

In high-end manufacturing, enterprises in the United States, Germany and other western developed countries have been the leaders in the fields of high-end equipment, engines and materials, thus earning high profits and controlling the entire manufacturing sectors. However, China's high-end manufacturing has struggled in last decades, and still lags far behind. A thorough analysis revealed that most of the high-end manufacturers in the United States, Germany, Japan and other western developed countries are well established with a hundred years of accumulation in innovations, core technologies, industrial standards, data and knowledge. This accompanies with a massive amount of investment, i.e. tens of billions of dollars. It seems very difficult for late comers to penetrate into the high-end manufacturing.

Some people hold the viewpoint that the US manufacturing is becoming increasingly weaker due to the large-scale outsourcing of processing and manufacturing capabilities. However, such argument is not convincing. In terms of revenue and value, the US manufacturing in fact has always occupied the high-end of the value chain. It can be argued that the US manufacturing industry always plays a critical role in the US and global economy. Indeed, the US manufacturing enterprises have always occupied the top-end of the manufacturing industry. They control the global manufacturing industry through the financial investment, service, industrial standards and large data, and core technologies of high-end manufacturing. As US enterprises are advancing to the industrial Internet industry, the US' control over the global manufacturing industry is becoming increasingly tighter and stronger, rather than weaker. This is supported by the fact that the Purchasing Managers' Index of US has been substantially higher than that of China.

So, it is inevitable that the US manufacturing, service and financial industries will reunite under new Internet economy after a long period of division. As aforementioned, the US production-oriented service industry and financial investment industries were originally diverted from the manufacturing industry. Today, these three main industries are implementing the re-integration process following the "+Internet" mode. Through advancing towards the industrial Internet industry, these industries commenced another round of transformation and upgrading.

It is worth noting that the manufacturing industries or economies of the United States and other western developed countries have experienced a natural development process including gestation, birth, growth and competition in the last century. As the industrialization process took a long period of time, the socialized division of labor and the specialized industrial development were given sufficient time. Therefore, dislike China, these developed countries don't have industrial structure problems in general. What they need to deal with is the industrial transformation and upgrading issues. Currently, the priority has been assigned by the manufacturing of United States and other western developed countries to complete the process of informationization as soon as possible, in order to accelerate the transformation to the industrial Internet industry.

The manufacturing industries in US and other western developed countries are well-supported by the industrialization, informationization and high-end manufacturing in earlier stages. As a result, the transformation and upgrading of their manufacturing industries to the industrial Internet will be relatively natural and smooth.

In terms of developing high-end manufacturing, latecomers will experience significant challenges or have little chance of catching up with the western leading enterprises. In fact, the key lies in higher investments and the integration of global resources. No matter how fast the industrial Internet is advancing, the high-end manufacturing and the control of core technologies have always been its core values. It can be foreseen that, with future development and the significant improvement in financing capacity, the Internet and industrial Internet companies will certainly turn their eyes to the profitable high-end manufacturing by making heavy investment in the research and development of core technologies. Through the integration of global superior resources, the Internet and industrial Internet companies will compete with or conduct M&A on those traditional high-end manufacturing enterprises that have a century of history. It is inevitable that the high-end manufacturing will attract an increasingly number of competitors and experience emerging changes. It will be inevitable that Internet and industrial Internet companies will advance into the high-end manufacturing industry. A large number of Internet and software companies now are constantly marching into real economic sectors such as automobile, digital home appliances, digital medical care, robots, energy and so on. These companies include: Apple, Google and Microsoft (United States); Alibaba, Tencent, MI and Baidu (China). This is evidenced by their gradual advancement into the traditional businesses of high-end manufacturing industries and by their intense competition with those traditional manufacturing enterprises.

2 How Was US Financial Investment Industry Created?

The US manufacturing has sustained substantial growth in the last century, namely the industrialization process. During this development process, the manufacturing industry has been accompanied by the financial investment industry since its birth. Both industries have been experiencing prosperity and depression simultaneously. It can be concluded that the US manufacturing industry has been aided by the financial investment industry, while the latter has been benefiting from the former. As aforementioned, a certain proportion of banks and investment companies are the derivatives of manufacturing. Originally, the founders of these banks and investment enterprises might have been owners of those manufacturing enterprises. After achieving massive profits, some of them gradually transformed themselves into bankers and investors because of their familiarity with manufacturing business. Due to the natural linkage with the manufacturing industry, their businesses are always specialized or industry-specific. In other words, most US banks and investment firms are specialized by serving specific fields. Such natural survival, growth and cooperation process has ensured the healthy development of the US manufacturing industry as well as the continued dominance of the US financial investment industry. Both industries have been collaborating with each other during the last century.

What remain unchanged are dedicated businesses, industries and fields. By contrast, changes are constantly taking place in methodologies and the continuous transformation. Through a thorough analysis, I found that in recent two decades, a growing number of US banks (represented by Citibank, JPMorgan Chase and Goldman Sachs) and US investment companies (represented by Buffett and Soros) have increasingly realized the importance of their customers and the analysis of their businesses. These US banks and investment companies have been transforming their core businesses from selling banks' own products to serving the customers and their business. They attach great importance to understanding customers' actual needs and place focuses on the support needed by customers in their business development. To this end, the US banks and investment companies have been injecting a large quantity of expenditure directly or indirectly to the development of software, network, big data and AI. The dependence of US financial investment industry on software, network, big data and AI is not accidental at all. In addition to addressing its own needs on information management, most of the developed software is associated with banking business and customer service. Indeed, a large amount of developed software is in fact directly linked to the operation, management and business of borrowers or investees. In other words, most banks and investment companies are specialized and generally focused on specific industries and fields they are familiar with. Apart from funding, they now also provide customers with necessary network and software tools, and big data-AI-services. This practice provides valuable benefits to customers and investees. In other wards, apart from obtaining funds, these customers and investees can continuously receive in-depth consultation services from the financial institutions on management and business, which ensures the success of enterprises and their investment projects. Similarly, such practice also allows bankers and investors to obtain customers' real-time data on operation, management and business development at any time. Then, through analyzing customers' data, the financial institutions can gain a better understanding on the status quo of customers in terms of the operation, management and business development. As a result, loan and investment risks will be significantly mitigated, even achieving so-called zero-risk investment. Perhaps this is the recipe for success mastered by financial magnates such as Buffett, Soros, Rogers and so on. In other words, the success of financial tycoons relies more on the powerful support that is gained from their huge investment on potent business tools such as informationization, software, network, big data and AI. With the strong support of information technology and tools, the so-called investment risk can almost be eliminated. This is a common practice adopted by successful financial tycoons such as Buffett, Soros and Rogers, which is the main reason they are trusted and admired by the public.

Aforementioned information technology means and supporting tools are easier to be said than to be practiced. This is exactly the massive gap between the financial industry of China and that of the US. Investments and loans made by the Chinese financial industry are significantly risky than those made by counterparts in US.

Most US financial investment industries are focused on specific industries and fields, being familiar with the businesses of investees and borrowers. They have close integration with information technologies such as software, network, big data and AI. It seems that a large number of US financial banks and investment companies have become leaders of the industrial Internet industry or have no difficulty in integrating themselves with the industrial Internet industry. Indeed, these are the strengths of US financial industry. This warrants deep thinking by the Chinese financial investment industry.

3 Joy and Sorrow Are Inseparable

Enterprises in the United States and other western developed countries have commenced the globalization process since the 1990s with the support of standardization, digitization, cyberization and big data. Driven by economic globalization, the leading manufacturers in the United States and other western developed countries have been outsourcing the medium and low-end activities (i.e. manufacturing, processing and services) in a bid to reduce their costs. Those medium and low-end manufacturing and service capacities and related technologies have been transferred to the rest of world, especially to enterprises in developing countries such as China and India.

Meanwhile, through constant layoffs and "downsizing", they are continuously moving to the high-end of the value chain. In addition, through transforming themselves towards the "asset-light" industrial Internet development mode, they place focuses on maintaining their monopoly position and competitive advantages to obtain enormous profits.

In fact, the manpower, scale, output value and sales revenue are not important factors from the perspectives of shareholders, industries and financial circles. Rather, more crucial factors are profits, dividends, high share prices and high efficiency on fund utilization. Therefore, it is inevitable that leading manufacturing enterprises in the United States and other western developed countries transform themselves to the "asset-light" industrial Internet companies during their development process. However, this has led to a series of issues in the United States and other western developed countries. These include: a large number of small to medium enterprises are gradually disappearing, a lot of employments opportunities are lost, a large number of industrial workers are unemployed, and the irreversible high labor costs. These issues have resulted in severe contradictions between entrepreneurs, financial capitalists and the governments on social policies. These contradictions are in fact very difficult to be solved.

In fact, those leading manufacturing enterprises, Internet companies and industrial Internet companies will not be able to survive and develop independently in industrial ecology without following critical factors: (1) the fundamental support from the large number of "asset-heavy" SMEs as the manufacturing base, and (2) the industrial support from the specialized and socialized industrial supply chain and service chain. Therefore, the tough task of facilitating leading enterprises to transform themselves to "asset-light" industrial Internet mode is emphasized in this book. In addition, this book highlights the significance impacts of "asset-heavy" SMEs on the industrial, economic and social development. It should be noted that the development routes of SMEs are not covered in the book.

With nearly 1.4 billion people, China is under enormous pressure on providing employment opportunities. China should take full advantage of rich labor resources and the large quantity of SMEs so that the vast number of "asset-heavy" specialized SMEs can collaborate with those "asset-light" leading manufacturing enterprises, Internet companies and industrial Internet companies. Consequently, a pattern of mutual support and coordinated development should be formed. This presents significant challenge to the governments at all levels.

There are some misunderstandings on the development of the Internet economy. Some claim that the Internet is a bubble that will eventually burst. Others claim that the Internet economy is misleading which should be treated with caution as it is costly but not effective. In my opinion, this represents a shallow understanding of the essence of the third industrial revolution, Internet and industrial Internet. It is a superficial understanding of the profound changes derived from this new economic theory to the economy, industry and businesses. Indeed, it is a misjudgment to the modes and trends of the global economic development.

In the US stock market, organizations and individuals all over the world are heavily investing in Internet and industrial Internet companies. Who will deceive himself/herself when he/she made genuine investment?

Meanwhile, a large number of enterprises, investors and stock investors have misperceptions of Internet companies and industrial Internet companies. They perceive that an enterprise will become an Internet company or an industrial Internet company which may appeal to the capital circle and even make their share prices surge if some achievements have been made. These include: the enterprise is exposed to software, informationization and the Internet; the enterprise has embraced some elements of Internet or has acquired some basic concepts from Internet (e.g. establishing a website or homepage, setting up an official WeChat account); or the enterprise has achieved partial informationization, online trading and online management through the application of some software. Actually, such way of thinking is very dangerous and may eventually make them feel extremely frustrated and may never be recovered. In fact, it is a system engineering with high level of technical difficulties, complexity and organization for an enterprise to become a genuine Internet company or an industry Internet company. The achievement of this goal requires heavy investment, the integration of superior resources and long-term efforts.

4 How to Find the Key for Opening the Door of Internet Economy?

Originally, standard is just a technological concept or category. Today, standard and the derived digital standard have become a key economic concept. All world-class enterprises are controlling standards and digital standards, no matter the traditional manufacturing, financial investment and service industries, or modern software and Internet industries. These include: manufacturing enterprises (e.g. Siemens, Boeing and GE), major banks (e.g. Citibank, J.P. Morgan and Goldman Sachs), investment firms (e.g. Buffett and Soros), service providers (e.g. McDonald's, KFC, Wal-Mart and Carrefour), software companies (e.g. Microsoft and SAP), and Internet industry companies (e.g. Apple, Google, Alibaba, Tencent and Baidu). In fact, all these companies have established a self-centered industrial technology system based on the standard and the digital standard, thus establishing a self-centered industrial ecosystem. In other words, first-class companies are making standards, second-class companies are making technologies, and third-class enterprises are making products.

In China, when you check in a hotel, you will often encounter such troubles, the temperature and humidity of room's air conditioning system are greatly varying and difficult to regulate and control; and you will often encounter difficulty in controlling the water temperature through the water mixing valve while bathing in the bathroom. You may also encounter those problems even when you check in a chain hotel run by Chinese companies. However, if you check in a brand hotel run by a firm of the US or other western developed countries, such as Holiday Inn, Sheraton and Westin, the room air temperature and humidity are very comfortable and easy to adjust, and the bathroom shower water regulating facilities are easy to control; even if you travel all over the world, these famous chain hotels will offer the consistent feelings to you. In order to do a good job in room air conditioner and shower water mixer, on the one hand, the support of key technology is needed; on the other hand, user companies should make the strict standards to the production enterprises, meanwhile, the standards must be implemented in practice. In fact, one of the main reasons why those hotel management companies in the United States and other western developed countries can win the favor of customers and stand out in the intense market competition is that they are the first-class standard companies and informationized companies who have established a huge industrial technology system and industrial ecosystem based on their industrial standards, thus dominating the world.

In China, the services provided by restaurants vary significantly. Even for the same brand, services will vary from one restaurant to another. All these restaurants have requirements and philosophy that are service-oriented, but there is lack of associated standards. Many restaurants in China are popular for the past century and are widely welcomed by consumers. However, it is a shame that most Chinese restaurants are unable to grow into large-sized businesses and carry out national or global expansion by adopting chain operation and industrialized operation. One of the main reasons is the inability to achieve standardized and informationized operations. As a successful model, Little Sheep Hot Pot not only has realized the chain operation and massive expansion throughout China, but also has established stores in the US. A careful analysis shows that Little Sheep Hot Pot is a restaurant that is specialized in selling mutton in hot pot. However, it has been designed and then has become a completely standardized and informationized enterprise from the very beginning, which leads to its success. In fact, at the global scale, all successful fast-food chains (e.g. McDonald's, KFC and Burger King) are the standardized and informationized businesses.

With regards to the manufacturing industry, all leading enterprises are the first-class standardized and informationized enterprises. These include: Toyota and Honda (Japan); Ford and General Motors (USA); Volkswagen, Mercedes-Benz and BMW (Germany). These leading manufacturing enterprises have achieved the standardization and informationization in product design, manufacturing, supply chain and enterprise resource management, as well as all services provided by 4S stores. Without standardization, it is impossible for the manufacturing industry to achieve the informationization. The concept of modern manufacturing industry won't exist without the standardization and informationization. Only the first-class industrial standard can facilitate the first-class manufacturing industry. Indeed, large-scale manufacturing must be standardized.

Now, let's have a look at the software industry, the core chip industry and the Internet industry.

Microsoft Corporation of the United States has developed an operating system software that is used by almost everyone in the world. The operating system is the software program that actually runs on the computer's CPU while all other application software are compiled to code 0 or 1 and carried out by the operating system.

Therefore, the capabilities and efficiency of operating system has become one of the core competencies of a computer. The operating system manages all hardware and software that are associated with the computer and network. Therefore, Microsoft has developed a powerful platform for the management of computer operations. Similarly, Microsoft has also developed a mighty platform to facilitate the development of application software, taking into consideration of all possible operations in the world. The scale of workload and the amount of code of the Microsoft's operating system are massive, which is beyond the imagination of ordinary people. So, how can Microsoft's operating system define and manage such a complex "ecology"? One of the key factors is the semantic definition as well as defining the corresponding logical relationships. Thus, rules, norms and logic are especially important. To put it simply, the most basic layer of the operating system requires a set of digital standards that are very complete, precise, and highly operational. In other words, the operating system is exactly the standard. Of course, the operating system is one of the core values of the IT industry, the Internet industry and the modern economy. Based on standards, Microsoft has established an enormous industrial technology system and industrial ecosystem, thus dominating the world.

Intel Corporation of the United States has developed general-purpose semiconductor CPU chips that are applicable to most of computers and servers in the world. All application software in the world has to rely on the operating system software to eventually perform physical computing on the CPU chip. General-purpose CPU chip is a key and core node for the application service and operation of computers, and of the Internet particularly. Therefore, its capacity and efficiency have become one of the core competencies of computers and the Internet. The design, operation, management and manufacturing techniques of CPU chips are extremely complex and beyond the imagination of ordinary people. If a CPU chip is magnified, a general-purpose semiconductor CPU chip made by Intel today could be much more complex than the three-dimensional images of the entire city of New York, Paris, Beijing and Shanghai. For example, a CPU chip could cover every aspect that needs to be covered by the complex urban design such as main roads, branch roads, various signal lights, pipelines, rivers, bridges, houses, buildings, stores, supermarkets, barber shops, parking lots, schools, government agencies, etc. Simply speaking, a CPU chip will cover all facilities and activities necessary for a city, covering a very complex "ecological" environment. So, where should we start when designing and developing a general-purpose CPU chip? One of the major key factors is to match the chip with the operating system software by making semantic definitions and logical definitions for everything and all activities in the world. Thus, rules, norms and logic are especially important. To put it simply, a complete set of precisely defined and highly operational digital standards must be established at the fundamental layer of general-purpose CPU chips. In other words, the general-purpose CPU chips are exactly the standards. Of course, the CPU chip is one of the core values of IT industry, Internet industry and modern economy. Based on its industrial standards, Intel has established a massive industrial technology system and industrial ecosystem, thus dominating the world.

SAP (Germany) has developed SAP-ERP, the well-known enterprise resource management software for the resource management of manufacturing companies, which is highly competitive in the world market. Manufacturing enterprises are affiliated to various industries. As a result, descriptions of enterprise resource management vary significantly according to the industry. Their semantic definitions and logical relationships will become extremely complicated. At the same time, even for enterprises engaged in the same industry, the description and logic of enterprise resource management will vary due to differences in scale, equipment, facilities, conditions, management methods, and operation modes. It is impossible to directly develop its enterprise resource management software using the software development platform provided by Microsoft operating system. Therefore, the core value and key in enterprise resource management software development is to build a so-called almighty platform through a high degree of abstraction and modular method. As a consequence, the high-efficient, low-cost and personalized development can be achieved according to customer's actual needs. Therefore, SAP has designed and developed a dedicated computer language for the development of enterprise resource management software. As a result, SAP has become the leader in enterprise resource management software. Maybe it is difficult to understand the word of "language". Indeed, such a "language" is actually the standard for enterprise resource management, which reflects the core value of the industry.

Starting from the digital libraries, Google (USA) and Baidu (China) are providing search engine services that are needed and desired by everyone. There are massive amount of books and materials at the national level and global level. For Google and Baidu, one of the major challenges is to classify the tremendous amount and different categories of books and materials, and to clarify their corresponding logical relationships. As a result, users can accurately locate and view the desired search results in the shortest time by using the search engine. This is a massive and sophisticated project. Today, that Google and Baidu are making progress and moving towards smart cities and smart earth by providing a wide range of services and applications such as Google-earth, automobile navigation, language interaction, language translation, and meal ordering services. It is difficult to live without these services today. However, a number of questions emerge: (1) how to define an enormous digital library? (2) how to define such large city? (3) how to define a more massive and humongous earth? and (4) how to determine the logical relationship and associations in a highly complex classification based on the above definition? Thus, rules, norms, and logic are especially important. To put it simply, a set of digital standards that define very precise and powerful operations must be established at the most fundamental layer of search engines provided by Google and Baidu. In other words, search engines are the standard. Of course, search engines today are one of the core values of the IT industry, the Internet industry and the economy. Based on the standards, Google has established a massive industrial technology system and industrial ecological chain, thus dominating the world.

In conclusion, the standards and digital standard are not only the foundations of manufacturing, financial investment, service, software and Internet industries, but also the foundations of realizing the digitization, cyberization, big data and AI.

Moreover, the standards and digital standard can serve as a key methodology in the construction of industrial technology system and industrial ecosystem by leading enterprises. At the deeper level, the standards and digital standards could be the function as a crucial "key" in realizing the industrialization, informationization and internetization. Indeed, they serve as a stepping stone to the development of Internet economy.

Compared with enterprises in the United States and other western developed countries, most Chinese enterprises have overlooked the strategic significance and core values of the standards and digital standards. This has become a major barrier to the strategic development of industries and the Internet economy in China.

Chapter 6
Discovering the Essence Through the Phenomenon, What Great Changes Have Taken Place in the Industrial Values?

Through the analysis of a large number of typical cases, the following questions will be illustrated by the author: Why the Internet is a new economic theory? What are driving forces of the "deindustrialization" and "asset-light" development mode? Why the manufacturing, financial investment and service industries undergo major changes? Why the core values of industries and economy will flow to the Internet industry, software industry, chip industry and high-end manufacturing industry? The core values of Internet economy will be clarified in this chapter.

1 Times Have Changed, Industrial Core Values Have Flown to the Industrial Internet

The author acknowledges that a significant proportion of people engaged in the manufacturing industry are still questioning the critical role played by Internet and industrial Internet. This is because they claim that clothes, household appliances, automobiles, airplanes, construction machinery, cement, iron and steel, and so forth must be eventually manufactured by the manufacturing enterprises and workers. Therefore, from their perspectives, the products and their businesses of rigid demands will always be assigned as the top priority, which is the core values of economy. Therefore, no matter how strongly the importance of industrial Internet has been emphasized, those people will still regard the software, Internet, 5G, big data and AI as supporting tools and as the complementary. This seems to comply with the economic principle: no matter how diversified the Internet companies, industrial Internet companies and their modes are, the fundamentals of their management and operations lie in the business and products of rigid market demands. It warrants attention that such perception might be appropriate in the past. However, according to the new theory of Internet economy, such perceptions have distorted the core values of today's economy and industries.

J. Yuchi, *Value Proposition*,
https://doi.org/10.1007/978-981-15-5025-6_6

The new China was founded in 1949. In the following 50 years, China experienced shortage of product supply due to the profound and constant influences of the planned economy. As a result, China's manufacturing industry, manufacturing products and industrial workers have remained in the dominant position in the economy and society. This has resulted in an ineradicable manufacturing-industry-focused economic value.

Let's take the iron and steel industry as an example. In 1978, in the early days of China's reform and opening up, a large supply of iron and steel products was needed due to the rapid economic development in all industries and sectors. In addition, China's own steel and iron industry at that time was not able to meet the needs of economic development. This resulted in importing a large amount of iron and steel. Meanwhile, China was desperately short of foreign exchanges, which became the bottleneck of importing iron and steel. To solve this problem, the Chinese leader, Comrade Deng Xiaoping, made a tough decision to establish a modern iron and steel enterprise in Shanghai. Resources were drawn from all over the country to introduce a complete set of steel production technologies and equipment from Japan, in spite of extreme financial difficulty at that moment. Consequently, Shanghai Baoshan Iron and Steel Co., Ltd was established to start mass production of iron and steel products that were urgently needed by the market. This laid the foundation for the "four modernizations" in China. Therefore, at that time, the steel companies and steel products were undoubtedly the core of economic values. Shanghai Baoshan Iron & Steel Co., Ltd. has always been a representative enterprise of China's steel industry. Through continuous innovation of steel models and processes, it has played a typical and exemplary role in the technological advance of the national steel industry. As a result, China's steel industry has achieved steady progress on maturity. Today, China has become the biggest steel producer in the world, far ahead of other countries. With the demonstration and driving effects of Shanghai Baoshan Iron and Steel Co., Ltd, it is no longer difficult to develop and produce ordinary steel products now in China. For instance, there are a large number of steel plants and enterprises in Hebei province which surrounds Beijing. Next to Shanghai Baoshan Iron and Steel Co., Ltd, there is a private large-scale steel enterprise in Zhangjiagang city.

At present, due to the significant overproduction of ordinary steel products, those steel enterprises' products are as cheap as the Chinese cabbages and radishes, causing heavy losses in the whole steel industry. There was a time when the steel industry was so vigorous and prosperous. Today, it is extremely difficult for the iron and steel enterprises to acquire bank loans, forming a sharp contrast with the past. Therefore, the state and industry policy makers today need to re-consider whether so many "asset-heavy" state-owned steel enterprises should be maintained. A large number of private enterprises have entered the steel industry while the development and production of ordinary steel products have become a largely pervasive activity in the society. Comparing the current situation with that of three decades ago when the Shanghai Baoshan Iron and Steel Co., Ltd was founded, is it still appropriate for the state to bear the risks of the steel industry? For the leading iron and steel enterprises like Shanghai Baoshan Iron and Steel Co., Ltd, we need to reflect on following questions: what are the core values of iron and steel industry today? Are they still the steel plants and steel productions? The fact that the share price of Shanghai Baoshan Iron and Steel Co., Ltd is an order of magnitude lower than that Shanghai Baoxin (a software and information service company derived from Shanghai Baoshan Iron and Steel Co., Ltd) adequately explains the nature of that problem.

Let's take the household appliance industry as an example. In the 1970s and 1980s, those televisions, refrigerators and other household appliances were luxury goods that were in a deficient supply in China. As a result, the ordinary Chinese families could not afford them, neither could they have access to these products. In late 1980s, with continuous improvements of people's living standards, the Chinese ordinary families would feel tremendously lucky and happy when they could buy a television set, a refrigerator or other household appliances. Meanwhile, very few Chinese people had opportunities to travel to overseas. If some people got chances to travel to overseas, from so-called duty-free shops, they were able to purchase imported television set, refrigerator, video recorder, washing machine or other home appliance produced by Japanese firms such as Sony, Panasonic, Toshiba and Sanyo. This would be enviable by others. Later, by introducing technologies and complete sets of production equipment from Japan and Germany, China's Peony, Wanbao and other household appliance enterprises began their production of televisions, refrigerators, washing machines and other home appliances. In late 1980s and early 1990s, the Chinese citizens were fanatically buying household appliances. The televisions, refrigerators, washing machines and other appliances of whatever brands were often out of stock. No matter which brand, they were in short supply and special means or "Tickets" were required to secure these products. This would be an indelible memory for the people of that era. During that era, the household appliances and the production enterprises were the core values of the society and economy. Soon afterwards, the household appliance industry experienced a leap-forward development with the emergence of a large amount of home appliance enterprises such as Haier, Chunlan, Midea and Changhong, etc. These enterprises continued to make progress and provide marketable products to the market, which greatly met the needs of people's lives. Meanwhile, as their products entered the international market, they had gradually become the internationally influential household enterprises. Indeed, household appliance industry was once the dominant industry in the Chinese economy.

Today, due to the relatively low technical threshold of household appliance industry, a large amount of enterprises and funds have swarmed into the household appliance industry, resulting in excessive competitions and oversupply of products. Consequently, the profit of a household appliance product averages as low as a hundred RMB. With low stock price, these household appliance enterprises are increasingly not favored by investors. It seemed that the major social function of home appliance industry lies in providing a platform for solving employment issues. Simply put, the household appliance industry is no longer profitable. In fact, a careful analysis reveals that the household appliance industry can still create profits and values. However, it is worth noting that its profits and values have mainly flown to the providers of complete-set technology and equipment for household appliance production lines, especially to those core chips providers. The Chinese household appliance industry was once in short supply where the demand far exceeded the supply. Today, its profit is too low to maintain operations.

Another interesting case is the development of construction machinery industry. In China, with the reform and opening up in 1978, all industries and businesses needed to be revitalized and rejuvenated. As a result, strong infrastructure was required to support the development of national economy. In the following more than three decades, a huge number of roads, railways, airports, bridges, ports and other infrastructure have been constructed. In addition, countless buildings and houses have been constructed to fulfill the demand of the state, society, businesses and citizens. Indeed, the entire China was a large construction site. As a result, the construction machinery industry and enterprises gained a rapid growth. These construction machinery enterprises continued to expand their market throughout the country, constantly supplying all kinds of construction machinery products to the market in large quantities. Thus, the demands from the construction market were greatly satisfied. Up to now, the scales and output values of these enterprises have been in the cutting-edge in the world. However, due to the low technical threshold of construction machinery industry, an increasingly number of enterprises and funds are entering the construction machinery industry. This has resulted in excessive competitions and oversupply. Eventually, enterprises in the construction machinery industry are suffering from declining profits and performances.

The fundamental reason is that the current construction machinery enterprises are lack of core competency. The core parts and components heavily rely on importing from overseas. As a consequence, 80% of the profits of the industry were distributed to the overseas countries and their leading enterprises. By contrast, most of the Chinese construction machinery enterprises concentrate on businesses that are not core values of the industry, e.g. processing, housings manufacturing and assembling. Therefore, various problems emerged when the macro economy slows down in recent years. These enterprises are struggling today due to the weak market and insufficient demand.

In fact, at the global scale, the core values of iron and steel, household appliances and construction machinery industries no longer lie in the processing, production, assembly and manufacturing. Rather, their core values are in the control of following key aspects: (1) core technologies in high-end steel types, innovations in household appliances and core chips, as well as the high-end core components of the construction machinery; (2) the control of industry standards; (3) the control of product's innovative design software, processing software, resource management software, life-cycle service software; (4) the analysis of big data generated by various software on product development, processing, manufacturing, management and service; (5) the capacity of building their own industrial technical systems and ecosystems. Following these principles, it can be considered that China's leading enterprises (e.g. Baosteel, Haier, Midea, Sany, Zoomlion and XCMG) should accelerate the process of standardization, digitization, network orientation, big data orientation and intelligentization. Consequently, they can transform and upgrade themselves into "asset-light" industrial Internet enterprises by shifting from the traditional "asset-heavy" development mode, in a bid to operate the industrial core values through the

"+Internet" development mode. Based on such platforms, these leading enterprises should make themselves available to all associated firms and businesses. As a result, a new industrial ecosystem will be created where Baosteel, Haier, Midea, Sany, Zoomlion, XCMG and other Chinese leading enterprises are the cores. Meanwhile, these leading upstream enterprises can achieve win-win solution and risk-sharing with those downstream enterprises. Thus, these leading enterprises could gradually open to the world to participate in the international competition. Fortunately, Zhang Ruimin, Chinese Haier's CEO has sensed the profound impacts derived from the new development mode of Internet economy on the household appliance industry. He has been actively exploring the transformation of Haier.

Times have changed. "Asset-light" and "Deindustrialization" are emerging. In the past, those iron and steel, household appliances and construction machinery giants were in the front-end of production. On the contrary, the industrial giants are now in the back-end by controlling the industry. As a result, the industrial core values have flown to the back-end, and to the "asset-light" industrial Internet giants.

2 Evolving Development Modes: Values Are Flowing to the "Asset-Light" Development Mode

We have discussed China's household appliance industry in the previous section. From a global perspective, China's household appliance enterprises and industry have experienced drastic fluctuations and vicissitudes. Meanwhile, internationally, the once-popular Japanese household appliance giants (e.g. Sony, Panasonic, Toshiba, Sharp and Sanyo) are struggling. A thorough analysis reveals that these Japanese household appliance giants have initiated the globalization process while ago. However, their industrial layouts and expansions in China, Southeast Asia and elsewhere were basically following the traditional development mode, i.e. the "asset-heavy" mode, rather than the new "asset-light" mode. In cases of excessive competitions, oversupplies and falling market demands, their product profits would continuously decline and these household appliance giants have to take all investment risks on their own. This has resulted in an industrial structure problem which is hard to be resolved in short term.

Times have changed. Despite the possession of advanced technologies, the traditional development modes no longer work in these household appliance giants. The fact is, Japanese household appliance enterprises (e.g. Sony, Panasonic, Toshiba, Sharp and Sanyo) are suffering massive losses year after year, accompanied by falling share prices and shrinking market values. This clearly reflected the significant impact of new "asset-light" development mode (i.e. industrial Internet development mode).

On the contrary, Apple company serves as a typical case of new "asset-light" development mode (i.e. industrial Internet development mode). Apple places more focuses on the control of innovative product design. An open approach is adopted in its product design through the Internet. Focus is placed on users' experiences with particular attention to the perceptions of Apple fans. Meanwhile, Apple has established a global industrial ecosystem by focusing on the control of following areas: key technologies such as designing and development of core chips; industrial standards and industrial technological systems; software development, e.g. the establishment of a powerful software development platform that is open to its industrial technological system and industrial ecosystem for joint efforts, co-creation, reciprocity and mutual benefits; control and analysis of big data for in-depth business consulting; financial services based on the aforementioned capacities, and investors' interests. A global industrial ecosystem has been developed through Internet. Apple's digital household appliance products with Internet concepts have constantly influenced the traditional household appliance industry and associated products. To manage the core values of its industrial ecosystem, Apple has outsourced all manufacturing processes and low-value services. By adopting new "asset-light" industrial Internet development mode, Apple has optimized its global resources allocation for the manufacturing, testing and services of its products, all of which are undertaken by its suppliers. As a result, Apple has achieved ultra-high profits and its investors have gathered super-high returns. More than 90% of the industry's core values have been distributed to Apple, its shareholders, and its investors. This fully reflected the law of industrial value. In case of market slowdown, its investment and management risks will be shared by all enterprises within the entire ecosystem. Meanwhile, Apple's market withdrawal mechanisms and risk control capacities (e.g. operation and investment) have been fully recognized. There is great potential for this new development mode. In fact, Apple is extremely familiar with the new theory of Internet economy and leading the new industrial development mode of Internet economy. Consequently, Apple has gained massive popularity in the global capital community, and its market value has reached over US 800 billion.

As shown in Table A.7 of Appendix A, the sum of current market values of those once-powerful Japanese home appliance enterprises (e.g. Sony, Toshiba and Panasonic) is equivalent to only a minor fraction of that of Apple company. There is clear contrast between Apple and those once-powerful Japanese home appliance enterprises. Apple is adopting new "deindustrialization" and "asset-light" industrial Internet development mode whereas those Japanese household appliance companies are adopting the traditional "asset-heavy" development mode. The performances of those Japanese household appliance companies in financial capital markets and stock markets are extremely poor.

Let's have a look at another typical case of industrial Internet development mode, i.e. a Chinese software enterprise that is marching into the real economy through adopting the "Internet+" mode. Lei Jun was once a famous entrepreneur in the software industry. Based on his profound knowledge of software, Lei Jun is now leading the MI company to soar in the Internet economy. Starting from smart phones, MI has built a powerful software development platform that is open to its industrial technology system and industrial ecosystem. Via sharing product design with MI fans on the Internet, emphasis is placed on the application experiences, joint developments, reciprocity and mutual benefits. Its industrial ecosystem is controlled through the industrial standards and industrial technical systems. On the contrary, all the production, manufacturing, assembling and testing activities are outsourced. Meanwhile, the production process is optimized through configuring global advantageous resources. Due to large production volume and low cost, its manufacturing costs are minimized. In addition, the networks and software are widely used to realize MI's internal and external management as well as service control. At the same time, its internal management is highly flattened by using a three-layer simple structure. Consequently, point-to-point and end-to-end management is achieved throughout the entire industrial ecosystem. There are a number of benefits such as: minimizing management costs and product prices, optimizing management efficiency and market reactivity, and maximizing profits for shareholders and investors.

MI places focuses on the analysis and control of big data, the control of in-depth business consulting, and consequently the control of financial services. Defining the core values of industrial ecosystem as the objectives, MI is adopting the new development mode, i.e. "asset-light" industrial Internet development mode. Founded just 6 years ago, by virtue of large outputs and low price, the sales volume of its smart phones ranks at the top level in both Chinese and global markets, posing a challenge to Apple and Samsung. In case of market slowdown, investment and operating risks of MI will be shared by all enterprises in its entire ecosystem. MI's mechanisms of market withdrawal and operation risks control is fully reflected. Therefore, MI's "asset-light" industrial Internet development mode has been fully recognized by the international financial capital community. In such short period of time, MI's market value is estimated to be more than 48billion US dollars, far exceeding that of the global renowned home appliances companies such as Sony and Panasonic. The development philosophy and development mode of MI are in line with the new mode and new theory of Internet economy. As a result, MI has accumulated abundant capitals in a short period of time. The investors and market are very optimistic about MI's future due to abundant opportunities. However, compared with Apple, MI is still falling behind in terms of the control of core technologies such as core chips. Therefore, MI has to make more efforts to catch up with the leading enterprises such as Apple.

Starting from smart phones, MI is changing the way how the smart phone industry and market operate in China. Meanwhile, MI is turning its attention to information appliances markets such as tablet, laptop, digital television, etc. This is made possible by relying on its advanced development mode, abundant capitals, strong capability on technological and management innovations. With a promising future, MI is most likely to lead the evolution of the digital and traditional household appliance industries and markets in China and beyond.

However, it is worth noting that MI's new "asset-light" industrial Internet development mode hasn't attracted an adequate attention from the industrial circles, financial capital communities, economic circles and various levels of governments in China. To many Chinese people, MI's mode is simply a special case. However, such a "deindustrialization" and "asset-light" industrial Internet development mode is actually the prevailing mode of future industrial and economic development. This indeed warrants deep reflection.

Indeed, the last decade has witnessed the evolution of the industrial development modes. Values are flowing to the new "asset-light" industrial Internet development mode. By contrast, the traditional "asset-heavy" development mode is falling far behind.

Let's take a look at the recent dispute between Vanke real-estate company and Baoneng investment company in China. What has happened between them? Opinions are entirely different on the Internet and in the society. Personally, I don't know Mr. Wang Shi, the leader of Vanke company, and has not communicated with him. However, I believe that Mr. Wang Shi's educational experience at Harvard University has significantly benefited him. In fact, it can be observed that Vanke's grand strategy is shifting towards the real estate industry-oriented Internet or industrial Internet. It has been a while that Vanke presented its transformation from the current development strategy (i.e. relying solely on themselves to buy land, build and sell houses) to the "asset-light" development mode (i.e. focusing on selling the brand, selling the mode, and operating the core values of its industrial ecosystem). It seems that Vanke company and Mr. Wang Shi profoundly learned and understood the essence and core values of the industrial Internet. In recent years, Vanke must have been vigorously developing standards, digital standards and software with orientation to the real estate industry. Vanke must have been studying the methodology of how to control and realize big data and its analysis in the real estate industry. Vanke must have been reconstructing its management systems, service systems, business processes and resources allocation based on cloud-based service and data center. As a result, the technical system and new industrial ecosystem are developed for the real estate industry with Vanke as the core. Meanwhile, the cooperation and integration with the financial investment industry is strengthened. With such capability in place, Vanke will inevitably merge a huge amount of real estate enterprises and resources in China in the future. In other words, faced with a difficult time and positions right now in the realm of Chin's real estate industry, the vast number of real estate companies can take advantage of Vanke's superior brand, technologies and management to finance, to acquire land and other resources, and to conduct real estate business independently provided they: (1) are willing to "join" Vanke's industrial ecosystem; (2) conform to Vanke's industrial standards; (3) use Vanke's management & service information system and software; (4) purchase Vanke's management, consulting and services; and (5) purchase high-quality and low-cost supply chain products provided by Vanke. Consequently, the core enterprise of Vanke and those subcontracting enterprises will achieve numerous benefits such as: the highest efficiency, profit maximization, cost minimization, and the sharing of benefits and risks. It can be foreseen that according to this new industry development model, Vanke and its brand will most likely occupy more than half of China's real estate industry in the future. By realizing the core values of the real estate industry, Vanke will become an industrial Internet enterprise in the Chinese real estate industry. Eventually, with strong competitiveness, Vanke will likely become a financial service provider and the genuine hegemon in the Chinese real estate industry. It is no wonder that Baoneng investment company has spent tens of billions of RMB to compete for Vanke's equity. Of course, there is nothing wrong with Baoneng's acute vision. Furthermore, Vanke deserves applause and praise for its transformation strategy towards the "asset-light" industrial Internet enterprise of the real estate industry as well as the new development mode.

Indeed, the development mode is constantly evolving. In the past, those real estate giants were in the frontier of construction industry. In the future, real estate giants will be the "asset-light" and stay in the backstage to control the industry. The core values of real estate industry will flow to the backstage and the new "asset-light" industrial Internet development mode.

3 Different Hardware, Different Prospects; Values Flow to Chip Industry and Software Industry

In the Chinese traditional manufacturing and IT hardware industries, a similar phenomenon re-occur constantly in the process of industrial development. Let's take television, refrigerator, automobile, computer and other typical products as example. Initially, these products were luxury goods that had to be imported from overseas. Due to the popularity, these products were often in shortage. Later, domestic enterprises began to introduce production lines and manufacture products with their own brand, which were best sellers with relatively higher prices and profits. This industry seemed to have very prosperous future. Then, due to the low technical threshold of these products, a large amount of enterprises and capitals have entered the industry. This has resulted in excessive competitions, poor markets, declining prices, low corporate profits, which indicated a struggling industry. In China, such industrial and economic phenomena were very common. Today, the profit of a television, a refrigerator or a computer is as low as 100 ~ 200 Yuan, while that of a car is merely about 2,000 Yuan. This is because these enterprises are generally lack of technological innovations, core technologies and added values of their products. From the perspective of the law of value, their work is mainly concentrated on designing a shell, either in plastic or steel. Thus, these products would be obviously lack of values, profits and market values. Therefore, if the technical threshold of a product or an industry is low, many enterprises will join the industry and the products would be lack of added value. As a result, all enterprises would mainly compete by means of lowering prices, which would inevitably cause the lack of value, profits and market values. Consequently, the industry will lose the favor of the capital community and fall into perplexity and degradation. In fact, China is still willing to classify some information technology hardware (e.g. desktop computers) as high-tech products. However, in essence, they are almost no different from hardware products of the traditional manufacturing industry.

So, where have the profits and values of the products (e.g. televisions, refrigerators, automobiles and computers) and their industries flown to? A careful analysis reveals that the main profits and values of the home appliance industry (e.g. TV sets and refrigerators) have flown to the chip industry. Meanwhile, the profits and values of automobiles and its industry have flown to the automotive electronics and chips industries whose prices are rising at a rate of more than 20% each year. As for a CNC machine tool, 90% of its value lies in NC chips and process software. Similarly, the core values of IT industrial hardware (e.g. computers and servers) lies in the core chips. Vast majority of market profits have flown to the core chip developers.

Intel and its general-purpose CPU chip products can serve as a most typical case. In brief, Intel assigns half of its developers and resources to the development of general-purpose CPU core chips. The structure of modern chips is very complex and beyond the general imagination of human beings. As a result, hardware manufacturers (e.g. computers and servers) are experiencing a growing level of difficulties to directly apply chips to the development of hardware products and systems. Thus, Intel has to allocate the other half of its developers and resources to the application development of CPU core chips, i.e. the R&D of various chip-based application solutions and associated key technologies. The inevitable consequence in a long term is that only innovation can create value. Therefore, the core chip developers will control the manufacturers and their profits of hardware systems (e.g. computers and servers). Meanwhile, those hardware manufacturers will no longer be required to conduct R&D of key and critical technologies. No innovation, no creation of values. As a result, those hardware manufacturers will gradually become attached to the core chip developers. They will gradually degrade into the manufacturers of hardware shell and traditional manufacturing businesses. Meanwhile, such hardware production industries will gradually shift from the United States and other Western developed countries to the developing countries. This will lead to increasing profits and values for the core chips developers; and decreasing profits and values for those hardware system manufacturers (e.g. computers and servers). In capital and stock markets, the valuations and share prices of core chip developers are extremely high while those of hardware system manufacturers are very low.

Table A.8 (Appendix A) presents the market value of Intel (supplier of general-purpose CPU chip) compared to typical manufacturers of computer and server equipment, which shows a massive gap. As a manufacturer of computers and servers, Lenovo is struggling with its, e.g. a huge loss in 2015.

The development mode adopted by Intel in the R&D, manufacturing and application service of general-purpose CPU core chips has gradually become popular in the chip industry. Therefore, this mode will have a profound impact on all the traditional manufacturing industries. This means that those traditional manufacturing enterprises (e.g. household appliances, automobiles and machine tools) will see their profits and values constantly flowing to the chip industry. In the era of industrial Internet, the traditional manufacturing industries will gradually concentrate on the processing, assembly and manufacturing of mechanical shell. As a consequence, their valuations and share values will continue to decline in capital and stock markets.

In fact, one fundamental in the R&D of manufacturing techniques and design of chip industry lies in the software development. Therefore, it is inevitable that the industry's value will flow to the software industry. Meanwhile, with the development of industrial Internet and Internet of Things, there is no doubt that the chip industry's development must provide support and cater to the Internet. Consequently, as cores of the network, various core chips will become key nodes of the Internet, industrial Internet and Internet of Things. This allows us to better understand the core values of chips. The core chips and chips-based televisions, refrigerators, automobiles, computers and servers seem to be hardware. However, different hardware, different prospects. Values are flowing to the chips industry, software industry and Internet industry. There are vast differences in the future between the traditional manufacturing industry and IT system hardware industry, and the chip industry.

In general, the R&D and manufacturing of core chips will become an integral part of high-end manufacturing, Internet industry, industrial Internet industry and Internet of things industry in the future. It will occupy the high-end of industrial value chain and the commanding height of industrial competition. In the future, the values and profits of those chip-based hardware system manufacturing industry will continue to fall, falling out of favor in capital and stock markets. Therefore, the traditional processing and manufacturing industries must become more specialized and consequently improve their added-value and core values through continuous innovations. Only this way can the traditional processing and manufacturing industries survive and sustain development.

Today, the capital communities in the United States and other Western developed countries have reached a consensus on the support and development of core chips, software, Internet and industrial Internet industries. This consensus will introduce a profound impact on the traditional manufacturing and IT system hardware industries. Furthermore, it will greatly affect the future of enterprise's operation and financing, and the stock market. China is a nation with an economic structure that focuses on the traditional processing and manufacturing industries. Our industry circles, capital circles, and economic circles must confront such severe reality and challenge.

4 Today's Software Is Different from that of Yesterday; Software Industry Has a Thriving and Prosperous Future

Originally, most software was point-wise and stand-alone. In China, the common perception is that software is a tool oriented towards management, product development and services. It is a tool that plays a supporting and auxiliary role, or even a dispensable role. Therefore, the society, industries and businesses have never attached a great importance to software's core values and its critical role in the industrial development. However, today's software is totally different from that of yesterday. Significant changes have taken place in today's software, e.g. the R&D methodologies, reflection of its core values as well as its impacts on various trades and industries. As a major core of industrial and economic values, today's software has a thriving and prosperous future.

From the perspective of the development methodology, today's software is generally based on a standardized, universal, open, structured, so-called modularized and "fragmented" platform. First, we must clearly determine and define an ecological environment of all businesses, management and services to be dealt with by the software applications. Second, we must define a set of fine-grained digital standards and logic, based on which we must develop tens of thousands of small modules that are application-oriented. Eventually, a complete and general-purpose software development platform will be formed. The challenge of such development methodology lies in heavy workload, high technical difficulty and long construction period for building such software development platform. To do so, software companies have to invest huge amounts of capital, manpower and resources. However, the advantage is that once the software development platform is in place, the market demand-driven application developer could simply use those corresponding modules. As a result, the application development would proceed as efficiently and rapidly as "building blocks". With high efficiency and flexibility, the application development would consume little time and money. Furthermore, by making most of the software development platform available to the public, such software platform company can quickly build its industrial ecosystem and industrial technical system based on its own platform and standards. This will attract an increasing number of downstream enterprises to participate in the entire ecosystem and realize their values. As a consequence, the software platform company could finally control and monopolize the market. Therefore, as long as a software company has built such a software development platform, it would possess core values and core competitiveness.

In terms of functions of the software application, the original product design software was simply a graphical design tool. However, today's design software has integrated with a huge amount of knowledge accumulated by human beings over many years in product design and practices. For instance, the Dassault's CATIA software is a typical representative of such software. In fact, this software serves as a platform for product design and development. The design and development of a real airplane could be finished in a shortest period time through the fast design of CATIA software and the constant optimization of simulation software. In the past, a firm's product design often relied on several designers or a team. Today, with Internet-based open software development platforms, enterprises can organize the best design talents across the nation and even the world to conduct a collaborative development online, thus sharing the wisdom and knowledge of mankind. At the same time, through solidifying such wisdom and knowledge accumulated over a long period time in the software development platform, a knowledge base for designing and development can be built to achieve the sharing of designing knowledge. This will make the product design increasingly simple, convenient, fast, efficient and intelligent. As a result, manufacturers can respond quickly to the market demands. However, the fact is that, on the one hand, such a knowledge-based design software will significantly facilitate the mechanical product design; On the other hand, it reveals that the core values of product design have gradually shifted from the traditional manufacturing industry to the specialized software industry.

Initially, machine tools were simply used to perform some basic and predefined machining operations. On the contrary, today's CNC machine tools could quickly and extensively process all types of complex products. Nevertheless, the key lies in the NC part rather than the mechanical part of a machine tool. In particular, the most significant difficulty and value lie in the process software of NC applications.

An enterprise's CNC machine tools are more likely to possess higher processing capacity and values if it owns a greater number and variety of process software, technology library and knowledge base. This in turn translates into more profits and values for the enterprise. Therefore, a major focus of today's CNC machine tool industry is the development of process software. Today, an excellent and high-value process software needs to integrate a large amount of knowledge and data accumulated by technical engineers and developers over a long period of time. The last decade witnessed the continuous development, improvement, optimization and evolution of process software. It can be foreseen that in the future, CNC machine tools may be provided to those processing enterprises free of charge. The incomes of CNC machining tool manufacturers mainly derive from services to the process software needed by machining process. This allows CNC machine tools processing enterprises to maximize their profits and the value of process software. It can be clearly observed that the core values of CNC machine tool industry have flown from the mechanical parts of machine tools to the process software, and to the software industry in specific field.

In the past, management software was simply a management tool. Today's management software is required to incorporate flattened and modern economic management ideas, philosophies, strategies, business and processes. As a result, a point-to-point, end-to-end, transparent, efficient and cost-effective network management can be achieved. With the point-to-point and end-to-end management of network and software in place, today's enterprises need fewer levels of management, fewer management agencies and less management personnel. Consequently, enterprises could significantly cut management staffs and reduce management costs, thus tremendously enhancing the management and decision-making efficiency. Business owners will have capacity to implement efficient, fast and accurate management through software, network as well as the analysis and reporting of data. Such management mode will enable enterprises to flourish.

Shuanghui Group is a well-known Chinese enterprise engaged in meat processing, production and services. Through powerful management software and information technologies, Shuanghui Group and its management can conduct a point-to-point, direct, high-efficient and delicacy management via smart-phones & Apps, e.g. the management of each pig of production bases, every meat store and every supermarket across the country. As a result, the intermediate links and management personnel of Shuanghui Group have been greatly simplified. This led to numerous benefits, e.g. significant reduction of management costs, improvement of management efficiency and profits. With the support of standardization, informationization, network orientation and big data, Shuanghui Group has begun to take over a prestigious US meat company. This relies not only on its abundant capital and courage, but also on the support of network, software, big data and outstanding capabilities on information management. With such capabilities, financial investment circles will be increasingly inclined and willing to invest in Shuanghui Group and jointly expand the production and services. In turn, Shuanghui Group is transformed into an industrial Internet company and a dominant leader in the meat industry. Today, the scale and business of Shuanghui Group have experienced continuous growth and expansion. In addition, Shuanghui Group has grown into a thriving business in meat industry. There is a paradigm shift of enterprise management, i.e. from "ruling by people" to the scientific management based on software and data. Values are flowing to the software, networks and big data.

In the past, service software was simply a service tool. By integrating business clients, service system and service mode with business model, today's service software has achieved timely, rapid and accurate services; personalized and customized services; point-to-point direct services; systematic and networked services, as well as product's life-cycle services. With such software and capabilities in place, enterprises can realize the shift of business model from yesterday's selling products to today's selling services throughout a product's life-cycle. The value of a product is maximized by taking a large proportion or even all profits throughout its life cycle.

In the past, the commercial operation of airlines has to through very complex processes. First, airlines (e.g. Air China, China Southern Airlines and China Eastern Airlines) purchase aircraft from Boeing or Airbus. Second, they have to acquire engines from Rolls-Royce or GE. Third, airlines request Boeing and Airbus to fit engines onto their aircraft. As an extremely sophisticated and complex mechanical product, the engine involves high-complexity business, technologies and personnel related to complex and precision machinery and machining, high-value and fine thermal component materials, electrical control, electronics, computers, chips and software. Hence, it presents a significant challenge to airlines to guarantee the daily management, operation, maintenance and monitoring of engines. This is especially the case for routine repairs, minor repairs, medium repairs and major services of engines. This problem has plagued the airline's operation and services for a long time. From the airlines' point of view, they are service providers that hope to start their operations with qualified pilots and flight attendants. In fact, airlines are not willing to spend a large amount of money, manpower and resources on their unfamiliar and costly multidimensional repairs and maintenance businesses that involve machinery, electrical system, materials, electronics, computers, software and so forth. So, who is appropriate for such complex repairs and maintenance works? The answer is very easy. It is more appropriate for those enterprises that design and manufacture aircraft and engines to provide the professional services such as the management, operation, maintenance and monitoring of aircraft and engines throughout their life-cycle, particularly the repairs. To response the airlines' concerns, Rolls Royce(UK) and GE(USA) have transformed the business model from selling engine products directly to airlines to leasing their engines to airlines during the entire life-cycle of engine products. Airlines are now charged by hours. In this new service model, airlines no longer have to engage in or care about the complex and costly businesses that involve machinery, electrical system, materials, electronics, computers and software. Rather, they just concentrate on the operation and services of aircraft. Meanwhile, Rolls Royce and GE can use network, software and real-time data to monitor the status of each engine across the world. Consequently, Rolls Royce and GE can charge airlines by operation hours and collect the full value of engine products throughout their life-cycle time. In addition, the engines have become much safer because of

Rolls Royce and GE's professional maintenance, monitoring and repairing services. Hence, the shift from selling products to selling services has helped both parties to engage in their respective professional services, thus achieving a win-win situation. By virtue of continuous innovation on technologies, organizational management and service model, today's Rolls Royce or GE is indeed an industrial Internet company in the field of high-performance engines, and a dominant leader in the engine industry. Meanwhile, as engines have become safer, aircraft in operation have become safer too. As a consequence, people are more willing to choose such a fast and safe means of transport. With better reputation of airlines, the credibility of the civil aviation industry will be further improved. This will finally create a prosperous civil aviation industry and an aeronautical manufacturing industry, thus achieving a multi-win outcome.

There is an enterprise named ShaanGu which is engaged in typical rotating machinery products in Xi'an, China. Through the software development and its application for complex mechanical products (e.g. remote monitoring, operation and maintenance of products), the enterprise has actively explored the pathway of industrial transformation from selling products to selling services. This deserves encouragement and publicity. It can be observed that, with service software in place, products' values have flown to services and the software.

There are four major links in manufacturing enterprises, i.e. design, manufacturing, management and services. With the capability to control these four major links, manufacturing enterprises will possess invincible market competitiveness, thus becoming a powerful enterprise with core values. It is a fair comment that today's industrial software is the fruit of human wisdom and knowledge, and core values of the manufacturing industry. The values of manufacturing industry has flown to the service, the software and the software industry. The "deindustrialization" is an indisputable fact.

Meanwhile, due to the popularity of Internet applications in recent years, the perceived values of software and application service continue to be recognized and affirmed by tens of thousands of online users. Software industry has gained a growing level of recognition in capital community and stock market as it brings convenient, fast and practical services to everyone. In the market, an increasing number of software companies are being acquired by the giants in the Internet industry, software industry and manufacturing industry. Some M&A cases by Internet companies during the last decade are shown in Table A.9 (Appendix A). An analysis of Table A.9 reveals those small software companies are often equipped with special skills that are indispensable in the Internet industry and industrial Internet industry.

As a top priority, application experience is achieved by software. This trend has become the mainstream value orientation in the Internet industry and industrial Internet industry. In fact, the Internet and the industrial Internet are built out of a variety of application-oriented software. These software is not only the foundation of Internet and industrial Internet, but also the key to the dynamism of applications and the core values of enterprise. Today's software is totally different from that of yesterday. With values flowing to the software industry and the Internet industry, the software industry has a thriving and prosperous future.

In the Internet economy, the core values and profits of software and its applications will become increasingly higher in the future, and the software's significance will continue to rise. As a consequence, software and applications will gain heightened interests from the capital and stock markets. It can be foreseen that the values of future software will be far higher than that of the hardware system. Today, traditional giants (e.g. Sony, Panasonic, Sharp, Toshiba, Hitachi, Nokia and Motorola) that take hardware products and systems as their primary businesses are suffering losses year after year. By contrast, the emerging companies whose businesses are dominated by the software, application services and Internet business are booming, e.g. Google, Apple, Microsoft, Samsung, Alibaba, Tencent, MI and Baidu. A fundamental of novel economics theory is that the core competitiveness of enterprises must reflect the core values of Internet economy today. It is necessary that such development trend is paid great attention from the China's manufacturing industry, software industry, Internet industry, financial industry, securities industry and the stock market.

5 Today's Products Are Different From That of Yesterday; Future Products Will Embrace Digitization, Internet Orientation and Intelligentization

In the past, products were basically simple and single-function. In fact, in addition to the basic functions, enterprises and individuals will also expect highest performance cost, best service, more values, best use and best functions from the purchased product. With continuous development and innovation in information technologies (e.g. chips, software, Internet, 5G, big data and AI), it is made possible to achieve such expectation. Today, digital, networked and intelligent products continue to surprise us. They bring about terrific feelings to the enterprises, the society and individuals. Meanwhile, tremendous changes have taken place in industrial and economic formats, leaving some people joyful while others sad. Therefore, entrepreneurs have to adapt themselves with the development of this era by constantly adjusting their strategies, actions and behaviors according to the changing situations.

Once upon a time, in the 1980s and the early 1990s, many Chinese people were extremely crazy about purchasing a TV set. They even resort to underhand means and plead for help everywhere. Today, this type of single-function televisions have become so obsolete that nobody shows any interest in them. In terms of forms, today's televisions have become the integration of monitors, computers, STB and networks. However, this is far from enough. Today's televisions requires the support of massive contents. Through televisions, users can do various things such as shopping, ordering foods, making hospital registrations, reading books, studying, playing video games. They can enjoy all excellent and convenient services. Today's televisions are different from that of yesterday. Contents and service experiences are the top priorities. Indeed, the value of the industry has flown to the chip industry, the software industry, the Internet industry and digital content providers.

Once upon a time, Nokia mobile phones were so popular that everyone was eager to secure one to keep up with the trend of the times. They had a dream of using mobile devices to change the world. It took Nokia only a few years to become a dominant leader in the world's telecommunication industry, which also became a leading industry of Finland. Both the state and people of Finland were very proud of Nokia.

However, Nokia mobile phones feature single-function, e.g. making phone calls and text messages. Nokia's prosperity did not last long. As circumstances are constantly changing with the passage of times, technological advancements and ideological innovations of human beings have never stopped. Steve Jobs is respected and admired by everyone across the world. He has carried out more profound innovations on technologies, production mode, Apps and service experience on the basis of Nokia mobile phones. As a result, Apple's smart phone (i.e. iPhone) emerged. The iPhone's functions and fashionable style have greatly changed the human beings' perceptions of the world and transformed the IT hardware and software industries. The explosive development of iPhone and iPad-based application software and service experiences have pointed out the direction of the industry, and triggered the real take-off of the Internet industry. In terms of forms, today's smart-phones have become the integration of mobile displays, communication device, computers and networks. However, this is far from enough. Today's smart-phones also provide powerful supports through massive contents and personalized services. Today, no matter where you are, as long as there is Internet access, you can enjoy everything with the terrific, convenient and fast services. These include: Internet surfing, sending and receiving e-mails, working, chatting, shopping, ordering food, making hospital registration, navigating, reading, studying, playing video games and so on.

Through iPhone, we can discover such a massive industrial ecosystem, so many entrepreneurial opportunities, so many technological inventions and creations, and so much industrial values are explored and realized. In addition, Apple's new "de-industrialization" and "asset-light" industrial Internet development mode has changed the understanding and cognition of human beings on the manufacturing industry. The traditional "asset-heavy" development mode of the manufacturing industry has been completely changed. Similarly, such mode has changed the value orientation of enterprises, industries and economies, making the development mode of traditional manufacturing industry a history. As a result, the global industrial, financial capital and economic communities began to study, to follow, and to worship Apple's new development mode. This has created an era. Indeed, humanity has entered a much more advanced era of Internet economy. Today's mobile phones are different from that of yesterday. The digitization, Internet orientation and intelligentization have radically changed our world. Contents and service experiences have become the top priorities. Indeed, the values of industry have flown to the chip industry, software industry, Internet industry and digital content providers.

In the past, each machine tool or sewing machine produced by manufacturing enterprises was an independent device that is designed to accomplish a specific function. Subsequently, each machine tool or sewing machine produced by manufacturing enterprises was equipped with NC devices. They were designed to apply different processing software to make the same piece of equipment become more efficient and faster. Thus, more designs can be produced and a wider range of products can be processed, which create more values. Therefore, the wide variety of processing software is the key because it could better reflect the values. Today, every single CNC machine tool or sewing machine produced by manufacturing enterprises needs to have a higher level of flexibility and efficiency. This will allow manufacturers to carry out serial and parallel jobs for different types of CNC machine tools and sewing machines at the same time. Meanwhile, a high-flexibility network controlled by software and data could be built to accomplish production tasks in a much more flexible and efficient manner. Therefore, today's CNC machine tools or sewing machines do not simply mean one set or several sets of stand-alone devices. Instead, they represent a complex network and services that consist of various CNC machine tools, sewing equipment, software and data. The key lies in software development, data collection and applications. In fact, a large proportion of values has transferred from those equipment to the processing software, management software, service software, network, large data and AI. Today's products are no longer that of yesterday. Nowadays CNC machine tools or sewing equipment are largely focused on the individualized and customized mass production based on the standardization, digitization, Internet orientation, big data and intelligentization.

Robot, as the name implies, represents an intelligent mechanical product compared to traditional counterparts. Today, in industrial productions, the working environment may turn out to be very harsh or dangerous due to the high intensity of continuous work and high requirements for consistency standards. This may result in physical damage to workers. Therefore, the manufacturers will substantially use manipulators, robotic arms and other industrial robots to replace workers in order to significantly improve the production efficiency and safety performance. For example, manipulators and robotic arms have been extensively used on injection molding machines in order to significantly increase productivity and protect workers from the pollution. In the welding production line of automobiles, multiple manipulators and robotic arms have been used to perform the collaborative welding work. This significantly increases productivity, improves the welding quality, and ensures the consistency of solder joints. When welding inside the ship, the working environment is very poor, e.g. extreme warm and humid in summer whereas extreme cold in winter. Therefore, robots can be used to replace workers to perform welding. Similarly, a large number of stacking robots are needed in tire production industry due to the massive weight and size of tires as well as the continuous work intensity. In a nuclear reactor, human beings are not allowed to enter to perform works. Therefore, the daily work can be finished by robots. The insertion and extraction of the nuclear fuel rod can only be done by the robot and the robot arm. The traditional CNC machine tools can only carry out cutting and processing little-by-little and layer-by-layer, which is featured with lower efficiency, longer time and higher costs. By contrast, a new type of CNC machine tool has emerged. At the beginning, it can quickly perform a large-extent and rough cutting like cutting tofu. At a certain point, fine blades will be used for thorough cutting. This new type of CNC machine tool is featured with higher efficiency, shorter time and lower cost. This is a highly intelligent CNC machine tool, which can be also called robot.

Next, let's talk about service robots. Both bionic human and bionic animals must be able to perform the following tasks: walking, checking road conditions, avoiding obstacles, identifying graphics and images. They can exchange and interact with human beings. They can understand languages, commands and feelings of human beings, and consequently make appropriate actions and reactions. They serve human beings or act as a real person or an animal. As a consequence, they bring service experiences and values to human beings.

It seems to be complex and difficult to either make a usable industrial robot, or to make a satisfactory service robot. It is not difficult to make a robot-shape shell. The real technical difficulty lies in the key components such as motors and reducers that are used in critical parts such as wrists, knuckles and joints of the robot. We must ensure these core parts and key positions can realize their functions in a flexible, high-speed, accurate and efficient manner. In addition, we must ensure the excellent consistency, durability and reliability. Another technical difficulty is that we must develop a lot of process software to control the movements and operations of these robots. Furthermore, we need to develop a large number of software on graphic and image recognition, voice interaction and human-thinking logic in order to manage various behaviors of these service robots. Thus, the accumulation and development of a large number of models, algorithms, databases and knowledge-bases are required which provides powerful support to the development of industrial robots and service robots. Today's products are no longer that of yesterday. Nowadays, the robot products are focused more on the digitization, software, Internet and 5G orientation, big data, intelligentization, personalization and real-time. Indeed, the values have flown to the high-end manufacturing industry, software industry, big data, AI and knowledge-base providers.

Since everyone is different in heights and body shapes, the suit is a highly personalized and customized product. In general, we purchase suits from the tailor's shops or suit shops. After choosing the fabric and taking measurements, customers must pay a high price and wait a long time to get the suit. Today, Red Collar, a Chinese company offers high-quality, fast, efficient and competitive-price services in China and US markets. After choosing the fabric and taking measurements in Red Collar's clothing stores, the consumers' data will be soon transmitted to the Red Collar's cloud. The consumers can immediately know the production time and delivery time of their suits, and are very satisfied with Red Collar's products and services. Red Collar's suits has become increasingly more competitive in the China and US markets due to the excellent services, lower prices and shorter delivery time. In fact, Red Collar has derived its competitiveness from: (1) its long-term development of knowledge-base for the suit product design, (2) the development and application of its Internet-based management and service software, and (3) mass production technologies and capacities that are featured with high level of complexity and customization. This is achieved by means of information technologies and software on the basis of highly personalized product needs. Therefore, compared with those traditional suits manufacturers, Red Collar could maximize values of the industry by virtue of its shorter delivery time, lower cost and higher efficiency. Red Collar's suits are different from those produced by traditional suits manufacturers. This type of suits are products of the digitization, Internet orientation, intelligentization, personalized and customized mass production. Indeed, the value has flown to the software, Internet and big data.

In the past, automobiles were typical mechanical structural products. Their structure and values mainly lie in engines, transmission and chassis. Later, an increasing number of cars began to use ECU and other automotive electronic control systems. This indicates that core values and innovations have been transferred to vehicle control electronics and vehicle-mounted electronics. Today, online automotive products represented by Google have become a heated topic. It is not hard to imagine that, in the future, a car will be a mobile computer, a server carrying a large number of sensors, a variety of application software and multiple communication functions. In the future, cars will become nodes of the sky-earth integrated networks. Future automobiles can not only achieve autopilot, but also instant communication, automatic navigation, route selection, automatic selection of service station, automatic diagnosis of faults, software fix and automatic selection of service stores. During driving, we can order foods, purchase movie tickets and pay the fees. Meanwhile, our car can broadcast audio books, newspapers and information so that we can keep in touch with the rest of the world all the way.

In addition, our car can automatically detect driver's health status and provide all the services we can imagine. In summary, we will feel relieved, secure and comfortable when purchasing, driving and using a car. It is necessary that the "asset-light" industrial Internet development mode is adopted in the production of Google's online automotive products in the future. Under such development mode, all productions and manufacturing processes will be outsourced to the best manufacturing enterprises and supply chains, thus achieving the optimal resource configuration across the world. When purchasing automotive products, Google will offer the most cost-effective services. Today's automobiles are different from that of yesterday. Nowadays, online automotive products place more emphasis on the standardization, digitization, Internet orientation, big data orientation, intelligentization, individualization and mass customized production. Its values are flowing to the chip industry, software industry, big data, AI and the Internet and 5G industry.

6 Unlike Dispersed and Unstructured Data, Big Data Generates Core Values

In fact, data is ubiquitous and exists in every aspect of production and life. The design, simulation, process, production, management and service of discrete manufacturing industries (e.g. automobiles, home appliances, machine tools and construction machinery) will generate a large amount of data. Similarly, the design, simulation, process, production, management and service of process industries (e.g. petroleum refining, petrochemicals, iron and steel production) will generate even more data. In addition, the controllers, sensors, instruments, drives and actuators of the automation equipment and facilities could generate a large quantity of data. The computers, software and networks will also produce large amounts of data. Meanwhile, the switch, routers and terminals will generate a lot of data, so as the human life and service activities. There will be links, association, logic and meanings between data. In order to make the large amount of data useful and valuable, it is necessary to make sure that data benefit all aspects of production and living activities. First, we must find scientific ways and methods to obtain and store the data. Second, we must identify the association and logical relationships between data. Finally, we can obtain the expected answers and conclusions that will serve as feedback to optimize the production activities and improve the social life. Thus, we will be able to create more efficient productions, much better life and more values. This problem-solving method has been commonly adopted by the humanity for thousands of years, which is referred to as the traditional method. In fact, to become a valuable asset of humanity and core value, each method must be extensively studied, explored and practiced by anthropologists and experts for many years. However, it should be noted that, a lot of money, manpower and material resources have to be invested to obtain such a method. In addition, it often takes several years and even decades. A difficult and complex task (e.g. the development of an aircraft engine, a high-value precision material, a core chip, or a piece of process software) could cost a huge amount of resources and time. Besides, it may end up with a failure.

Today, human beings have entered the era of computers, software, information and Internet. With the support of computers, software, Internet and those sensors, we can collect, acquire and store all or the vast majority of data related to an industry, a product, or a service in the first instance. Consequently we can classify and sort these data into different categories for further in-depth analysis and mining so that the association, logical relationships and law between data can be identified. Thereby we can infer the existence of a so-called scientific method or law. In other words, despite not knowing the existence of objective law beforehand, we can infer the presence of a scientific law or method through the in-depth statistical analysis of large amounts of data. This is a hot topic in today's computer and Internet communities, i.e. "big data". Through big data approach, the massive data become useful, thus creating true values and core values. Today, a large number of industrialists and academics do not fully agree that the presence of objective laws unknown to the human beings could be discovered by means of the big data analysis. However, in the Internet industry, the big data method has been popularized in full swing, gaining an increasing level of recognition in the Internet and the society. The fact is that the big data has played an increasingly important role in everything, which constantly creates values.

In fact, I believe that the methodology of traditional cognitive law is not contradictory to that of big data. Rather, these two methodologies are complementary to each other. Big data can help industrial and academic communities to quickly identify the objective existence of possible rules and methods, thus accelerating the discovery of objective laws by industrial and academic communities. As a result, theories can be linked to and derived from practices. The faster and better innovation will benefit the industry and the society.

In the Ocean University of China, Ding Xiangqian, a famous professor in computer science has been guiding Dr. Yang Ning (one of his students) to work on the R&D of tobacco formula and its process software for many years. The first step was to study the tobacco quality and character, proportioning, technologies and other aspects of the tobacco industry so that a physical model can be built. Then, they established a mathematical model accordingly which can be used on computers. Prof. Ding has also made a bold attempt to study why so many people are willing to smoke even though they are aware the detrimental effects of smoking. What are physiological factors and laws that lead to tobacco addiction? Therefore, he directed Dr. Yang Ning to develop a model of physiological dependence at the same time. It is extremely complex to blend the three models together in a single software platform as it consists of three-dimensional business, information technologies and physiology laws. Prof. Ding and Dr. Yang's team have made great efforts and conducted a lot of research. The team was even laughed by others due to the significant difficulty to develop the physiological dependency model. This software platform was finally completed a few years ago and can be used in practices. Since the taste of smokers is constantly changing, tobacco factories must develop new cigarettes to keep up with the times. Therefore, for the tobacco factory, the core value lies in the continuous development of new cigarettes to meet the constantly changing market demands. In the past, the development of a new cigarette is similar to that of a new wine. First, tobacco tasters must select different tobacco leaves and then mix them together using different proportioning. After numerous iteration of repeated tastings, a probably best-selling cigarette can be created and produced finally. Therefore, the tobacco tasters has become an indispensable resource for the tobacco factories. However, it is necessary that the selected tobacco leaves and proportioning are consistent with processes of the current industrial production lines. Indeed, the inconsistency between the new cigarette development and the production line has presented a significant challenges to the tobacco factory and industry. Nowadays these problems have been solved for cigarette factories through the software platform developed by Prof. Ding and Dr. Yang Ning. The novel method of new cigarette development is as follows: (1) according to the market demand for new cigarette tastes, those cigarette factories will perform corresponding clicks on the computer, (2) a tobacco formula will be created and repeatedly tasted by those tobacco tasters who will make a preliminary judgment, (3) based on this, the optimization will be undertaken by a computer while the proportioning of new cigarettes will be determined fairly soon after several rounds of optimization, (4) finally, the software platform will be used to control the production processes so that the cigarette factories can start mass production of cigarettes. With such technologies and capabilities in place, the vast majority of the Chinese cigarette factories are commissioning Prof. Ding's team to develop new cigarettes and process software on the software platform. In this process, all Chinese cigarette factories must pass on all kinds of data related to tobacco leaves and processes to Prof. Ding's team. This leads to an unpredictable outcome. All data associated with the vast majority of Chinese tobacco leaves, tobaccos and processes are stored in a humble server located at a room corner of the Ocean University of China, which is actually the big data of Chinese tobacco industry. In other words, the core values of Chinese tobacco industry have been quietly transferred from those famous cigarette factories to this software platform and the stored big data. This software platform and the stored big data have actually taken control of those famous Chinese cigarette factories and China's tobacco industry.

Let's continue this story of big data. Due to the market competition, these famous Chinese tobacco factories have been constantly expanding their businesses by means of merging a large number of small tobacco factories across China in recent years. However, there is an associated problem. Since those merged tobacco factories spread across the country, it presents a significant challenge to maintain the consistency of tobacco products for the production of the same cigarette. This problem can be also dealt with by using this software platform and the stored big data. By analyzing the data on the platform, causes and corresponding solutions of problems can be identified. Consequently, by adjusting relevant process data on the software platform, the production of those merging tobacco factories can be controlled. Indeed, through this platform and data analysis, the cigarette producers can solve various issues during the procurement of tobacco leaves (e.g. quality and pricing). This is in fact another critical issue associated with the tobacco industry as all cigarette factories want to purchase high-quality tobacco leaves. More specifically, a guidance can be provided for the cultivation of tobacco plants across the country by using this software platform and big data. This is because farmers, growers and local governments are all extremely eager to understand the most suitable farmlands for the cultivation of tobacco. As a result, better-quality tobacco leaves can be produced; and the maximum efficiency and values can be achieved. There is no doubt that the software and big data are the core values of this industry.

We can continue to discuss this story of big data. It is likely such tobacco software platform can be translated into others businesses. For instance, it can be used to address the formula and consistency issues in alcohol, beverage and food industries (e.g. beer, wine, liquor production) and the pharmaceutical industry (e.g. the preparation of the traditional Chinese medicine). If that is the case, how can we measure the effectiveness and values of big data for these industries and trades? There are so many benefits associated with software and big data that we will stick with them.

Let's discuss the big data in the household appliance industry. Household appliance industry is highly service-oriented that serves almost all families. Through many years' development, leading home appliance businesses (e.g. Haier, Midea, Hisense, Gree and Changhong) in China have sold countless products. Have they taken control the big data of home appliances users? The answer seems to be yes. Once upon a time, products of household electrical appliance enterprises were so popular, all of which were implemented by the manufacturers directly. Therefore, these enterprises certainly took control all data of products users, such as names, home addresses, payment methods, telephone numbers, e-mails and other contact information of users; as well as product related information such as types, production batch numbers and performance index. This allows the household appliance enterprises to provide delivery, installation, maintenance and other after-sales services.

Later, with the emergence of intermediary (e.g. Gome and Sunning) in China, a new way of purchasing home appliances has emerged. The general public went to these Gome and Sunning hypermarkets to buy home appliances. In fact, a major change has already taken place in this period, i.e. those information and data related to home appliances users were transferred to the hands of Gome, Suning and other hypermarkets. In recent years, with the emergence of e-commerce, another revolutionary change has taken place to the sales mode of home appliances. By virtue of lower prices and superior competitive edge, the e-commerce platforms have attracted an increasing number of household appliances buyers. Indeed, the on-line shopping has quickly gained popularity. At this point, there is another silent revolutionary change, i.e. the information and data of the home appliance buyers have been transferred to the hands of Alibaba, JD.Com and other e-businesses. After the two revolutionary changes, those household appliance manufacturers have to resort to the hypermarkets (e.g. Gome, Suning) and e-businesses (e.g. Alibaba and JD.Com) if they want to acquire their end users' information and data. The complete information and data sets of household appliances users (i.e. "big data" of the industry) are increasing controlled by Alibaba, JD.Com and other e-businesses. For vast majority of household appliance enterprises, it is not possible to fulfill their after-sales service commitments (e.g. active services, life-cycle services and personalized services) without the access of the big data of home appliance users. In addition, their sales channels and pricing of products as well as funds are controlled by those e-businesses (e.g. Alibaba and JD.Com) and hypermarkets (e.g. Gome and Suning). At this point, let's determine the value of home appliance businesses. Do you think the industrial form of today's home appliance enterprises is more like a pure processing industry? In other words, are those once-giant home appliance enterprises simply working for e-businesses and home appliance hypermarkets today? Without innovations in technologies, productive relation, production mode and industrial development model, the enterprise is bound to be short of added-value and profits. Similarly, its stock value and market value will decline. Therefore, it is no wonder that today's home appliance enterprises are struggling in the market. Who has the control over the core values of the home appliance industry in the future? The answer is quite obvious. They are in the hands of those e-businesses.

Next, let's look at the lack of innovations in home appliance product services from the perspective of Internet of Things and Big Data. Almost every family has purchased a variety of home appliances, each of which is generally designed to have multiple functions. However, do these household appliance enterprises have the ability, technologies and means to understand which functions of these products are frequently used and specifically how they are used through the big data analysis? The enterprises should understand parts and functions of their products that require a high level of quality and reliability. By understanding product features that are basically not used by customers, the manufacturing enterprises can save the production cost by omit such functions in subsequent design and manufacturing stages. In simple terms, an excellent household appliance enterprise must have the ability to monitor and analyze the big data related to functional usage of their products. Consequently, a closed-loop system is formed. As a result, the enterprise and its products (e.g. design, technologies, manufacturing, quality and energy consumption) can be continuously optimized. It is imperative to undertake such continuous optimization, improvement and innovation. As a consequence, the enterprise will be likely to continuously increase their values and sustain a competitive advantage.

Let's imagine that a comprehensive service platform on citizens' health, medical treatment, medical insurance and medicine is established in China. This platform is based on a unified standard, the Internet, 5G, software, big data and AI. All resources and data (e.g. hospitals, medical equipment, facilities, physicians, patients, health insurance companies and pharmaceutical enterprises) provide inputs into the platform. As a result, the big data sharing of medical industry could be achieved. Consequently, we could use big data methods to carry out an in-depth analysis of various types of data. This makes it possible to predict the health status, morbidity, drug use, treatment, recovery and life span of every citizen. In addition, common diseases in local areas could be foreseen so that doctor resources can be allocated and distributed across the country. Thus, the allocation of hospitals and resources across the national can be predicted. This provides essential guidance to the pharmaceutical enterprises in terms of medicine production and demand forecasts. This will also allow us to predict the demand for medical insurance funds. Consequently, a harmonious development of the whole society can be achieved.

Big data is different from dispersed and unstructured data. Big data could generate core values. As the values are flowing to the Internet industry, AI and software industry, the future of big data is immeasurable.

7 Different Kinds of Services: Values Flow to Innovative Internet Services

Let's discuss and analyze the aforementioned case of the Chinese restaurant of Little Sheep Hot Pot from the perspective of services. Little Sheep Hot Pot is a restaurant specialized in selling instant-boiled mutton, which has become a service-oriented chain operation enterprise. Since the planning phase, it has focused on capital operation on the basis of standardization, digitization and Internet orientation. Its service mode and business model are different from those of the traditional Chinese catering industry. Little Sheep Hot Pot concentrates on the industrial eco-chain management rather than the restaurant management. Its instant-boiled mutton only serves as a carrier for realizing the core values of industrial ecosystem through services. In fact, all successful fast-food chains in the global market (e.g. McDonald's, KFC and Burger King) are the standardized and informatized businesses that provide us with safe and sanitary food. Based on standards and informationization, these enterprises have established an enormous industrial technology system and industrial ecosystem, thus dominating the world.

McDonald's is taken as an example for a detailed and in-depth analysis. At the beginning, McDonald's was a small US food store which concentrated on selling

American-style buns, i.e. hamburgers. Main attractions to customers were quality, price and smiling services. Based on growing popularity among consumers, McDonald's began the expansion of its stores and created a brand advantage locally. Later, with the support of standardization and informationization, McDonald's initiated the transformation of service mode and business model to realize a chain operation. Until now, McDonald's has realized the numerous values of the industrial ecosystem by transforming the business model from selling hamburgers and drinks to managing the industrial eco-chain. Finally, the McDonald's Management Corporation was formed. As a result, McDonald's has gained constant growth and continued to expand its business throughout the United States and beyond. Nowadays, McDonald's has become a leading brand and trademark over the world. In fact, McDonald's is a typical "asset-light" industrial Internet company today in the fast food industry.

In the past, we predominately went to stores and shops to buy living necessities. These stores and shops were small and relatively independent. They were entities with monolithic structure and a group of employees including all types of management, planning, procurement, accounting and transportation personnel. This thus resulted in inevitable high administrative costs which would necessarily be distributed to all commodities. Meanwhile, due to the small volume of procurement, the purchased goods' prices would certainly be higher. Therefore, these stores and shops would inevitably sell commodities at relatively higher prices, which meant consumers would pay more. Today, these small stores and shops are still ubiquitous in Chinese communities and on the streets. Their main function is essentially a convenience store or shop. Later, the chain-operation supermarkets emerged. With a wide array of goods, supermarkets can fully meet people's needs and demands. More importantly, the commodity prices in supermarkets are relatively lower, and consumers can help themselves to select and purchase the goods. As a result, supermarkets have gained increasing popularity from the general public. Today, we have been increasingly inseparable from these supermarkets. This new service mode and business model is in fact a deep understanding of services, which requires strong supports from the standardization and informationization. The service mode and business model of traditional retail business place focuses on the commodity sales. By contrast, the supermarkets concentrate on operating the industrial eco-chain, and those goods are simply the carriers to realize the core values of their industrial ecosystems through services. In fact, all the successful chain supermarkets (e.g. WalMart and Carrefour) are the standardized and informatized businesses that allow us to shop with ease, security and pleasure. Based on standards and informationization, they have established an industrial technology system and a massive industrial ecosystem, thus dominating the world.

Next, Walmart will be taken as an example for analysis. With the support of standardization and informationization, Walmart's service mode and business model are characterized by the global chain-operation and industrial development. Walmart defines the goals as operating the industrial eco-chain and achieving the numerous values of industrial ecosystem. Finally, the Walmart Management Corporation is formed. In fact, Walmart is a typical "asset-light" industrial Internet company today in the supermarket industry.

From the perspective of business mode, for each product sold by McDonald's stores, and for each commodity sold by Walmart supermarket, the cashier's POS machine will automatically reduce the stock and pass the massage in real time over the network to McDonald's or Walmart network-controlled distribution center. When the stock of McDonald's store or Walmart supermarket is reduced to a defined critical point, the distribution center will automatically send out a production order to those manufacturing factories and a delivery order also to the logistics distribution firms. This automatic process is at a very high efficiency.

From the management perspective, apart from necessary service staffs and one or two so-called store or supermarket managers, there is basically no other management personnel in McDonald's store or Walmart supermarket. Thus the management cost is minimized and the efficiency is greatly enhanced.

From the service perspective, McDonald's and Walmart supermarket have realized the standardized services. By providing safe and reliable products and highly consistent services at low prices, they possess powerful market competitiveness.

From the perspective of realizing corporate values, McDonald's and Walmart are the global renowned brands. Therefore, McDonald's or Walmart's management company will charge each new franchisee high licensing fees for brand and intellectual property. Apart from the renovation cost paid to McDonald's or Walmart's management company for a unified image or logo at the opening of new McDonald's stores or Walmart supermarkets, those owners of franchised stores or supermarkets must also pay for all kinds of uniform kitchenware or production tools. In addition, the franchised stores or supermarkets owners need to pay McDonald's or Walmart management company for the food ingredients such as bread, meat pie, fries, drinks as well as the costs of a large number of consumer goods to meet the daily business needs. Due to the constantly increasing demands for products, McDonald's and Walmart management company could continue to put pressures on those suppliers of raw materials and consumer goods for lower prices. As a result, McDonald's and Walmart management company can dig out and achieve more profits and values. Therefore, one of major tasks of McDonald's and Walmart management companies is to develop more franchisees based on market surveys as well as the analysis of big data generated by daily production and services. By providing more and better services, the maximum values can be realized.

With so much profits and values flowing to the "asset-light" businesses such as McDonald's and Walmart, it is no wonder that Walmart ranks the first among the world's top 500 enterprises year after year. Software, networks and big data generate core values that flow to McDonald's and Walmart, and to the new services modes. Future's services will be the standardized, digitized, intelligentized, data-intensive and Internet-oriented.

Significant changes have taken place in the information search and travel services. In the past, it was a complex and expensive process to locate the information, especially for those in remote areas. For instance, a science and technology staff living in remote areas such as Yunnan province or Xinjiang Autonomous Region in China needed to locate some information, data and reports for his scientific research. He would need to visit the library or information center in the county. If he failed to find the desired information in the county library or information center, he would have to travel to the city library or information center. If he failed to find the desired information again, he would have to take buses to visit provincial-level libraries or information centers. If failed again, he would have no choice but take a train to visit the national libraries or information centers in Beijing. The travel from Yunnan or Xinjiang to Beijing needed the cost of not only train tickets but also hotels and meals. The travel would be truly time-consuming. In the past, it took more than 70 h to travel from Yunnan or Xinjiang to Beijing. As a result, a round trip might take half a month. The efficiency of scientific research was obviously very low. Nowadays, with the help of Internet, everyone can find the desired information and data through search engines (e.g. Google and Baidu) on computers or mobile phones in such short time, low cost and high efficiency that are truly beyond the imagination of those predecessors.

In the past, when we traveled overseas or to other places of China, it could be a big job to drive to the destination. Even you can rent a car or be capable of driving, you might have no idea of local roads and streets. The map is generally complex and confusing. Today, Google and Baidu's Mobile Maps can provide extremely accurate navigation, which completely solves that problem. In the past, if you had a plan to travel overseas by yourself and you only possessed a limited capacity on English or local language, you might get deterred and choose a team-travel instead. Today, with the help of Google or Baidu's mobile language or speech translation software, you can freely travel overseas and use your smart-phone to ask for directions. New service experiences could fully liberate your desires and aspirations. In fact, Google and Baidu are working to provide services such as digital library, smart city, and smart earth to meet your expectations. Once upon a time, it could be a difficult task to buy a train ticket or air ticket that even required employing personal relationships in China. Nowadays, with the rapid development of civil aviation industry and high-speed trains, we can easily buy a ticket by simply clicking the mouse on computers or tapping on our mobile phones. In terms of taxis services, it was meant to be a business between taxi drivers and passengers only, i.e. passengers pay the fees and drivers provide services. However, in China, there are some taxi management companies that charge a certain amount of the commission fees but do not provide any value-added services. This has indirectly raised the taxi costs and has severely distorted the law of value. In addition, the opaque administrative licensing has become a resource pursued by the taxi industry. As a result, the taxi industry is being widely criticized by the society and the general public. Meanwhile, the public image of local governments has been damaged. Today, the emergence of online taxi-hailing service companies such as Didi and Dida has rekindled our hope for the industry. It is expected that the value distortion in this industry can be effectively resolved to satisfy the society and the public. In fact, taxi riders simply want to find a driver to deliver themselves to their destinations safely, quickly and cheaply. What a simple wish it is! How can it be so complicated?! Today, this sort of Internet service companies are constantly making progresses, innovating and creating values. They bring increasingly more beneficial services and experiences to us. Today, the earth is like a village. The wonderful human life lies in your clicks on computers and taps on the mobile terminal devices.

Consequently, you can make this complex world to serve you in an orderly, flexible, fast, high-efficiency and low-cost manner, thus making your life increasingly better.

The second half of the 1980s and the first half of the 1990s in China were the best period of time for the home appliance enterprises such as Chinese Wanbao, Peony, Japanese Panasonic and Sony, whose products sales far exceeded the supply. Thus, those home appliance manufactures became a dominant economic force. Since then, the home appliance industry embraced an explosive growth and attracted a huge influx of investments and production enterprises into the industry, which resulted in excessive competitions. From the year of 2000, with the support of informationization, the chain service began to emerge which transformed the service mode of home appliance industry. Hypermarkets (e.g. Gome and Suning) began to dominate this industry. Due to the lower prices and better services, Gome and Suning hypermarkets became the first choice to purchase various brands of home appliances. Meanwhile, as Gome and Suning sold more and more sound-quality home appliances at lower prices to meet customer expectations, they constantly requested those household appliance enterprises to lower prices and improve the quality. Despite declining prices, the household appliance enterprises had to meet the requests of Gome and Suning because of their heavy reliance on the enormous ordering from Gome and Suning. As a consequence, the household appliance enterprises were getting less profitable and more heavily reliant on Gome and Suning. Indeed, hypermarkets (e.g. Gome and Suning) in China have in fact taken control of those household appliance enterprises and their profits. It is a valid statement that Gome and Suning actually revolutionized household appliance enterprises such as Wanbao, Chunlan, Changhong, Panasonic and Sony. Since 2010, the Internet service mode emerged, which is characterized by the Internet support, network marketing, direct-delivery logistics, and service experience as the top priority. At the beginning, Alibaba online services rose rapidly with irresistible and invincible forces. Later, JD.Com and other Internet service companies began to squeeze into this market. Today, people do not have to go to hypermarkets (e.g. Gome and Suning) to purchase home appliances. Rather, they can simply purchase the lowest-price and good-quality home appliances from online shopping malls that are operated by Alibaba and JD.Com. Once online payments are made, the goods will be directly delivered to customers' homes for installations. Today, online shopping has gained the momentum. It is a valid statement that e-businesses (e.g. Alibaba and JD.Com) in fact have revolutionized hypermarkets as well as those household appliance enterprises. In addition, e-businesses have put enormous pressures to large commercial centers or malls and their tenants of specialty stores. These commercial centers or malls were developed by the real estate businesses such as Wanda (China) with a huge investment. This is because customers now treat these specialty stores as fitting rooms where they would find their desired style and models. Then, they could shop the products through online shopping malls. Changes in the service market are really fast, remarkable and encouraging. It can be expected that in the future, e-businesses (e.g. Alibaba and JD.Com) will adopt a "deindustrialization" and "asset-light" industrial Internet mode to make each manufacturer of home appliance and consumer goods to become a specialized and processing plant which just provides production activities for them. As a consequence, these e-businesses can control most of profits and core values in the home appliance and consumer goods industries.

In fact, MI, another industrial Internet giant must also be paying attention to this industry and market. Consequently, the majority of household appliance enterprises and consumer goods manufacturers will become purely processing enterprises, suffering from low profits and share prices. Nowadays, Internet service companies (e.g. Alibaba, JD.Com and MI) are increasingly taking the initiatives, controlling the fate of manufacturers of the household appliances and consumer goods. The home appliance enterprises, home appliance hypermarkets, large commercial centers, e-businesses are all offering services to the public. However, these services are entirely different. Software, Internet and big data generate core values. These core values are flowing to the innovative services, to e-businesses, and to the software industry, big data and Internet industry.

There is another typical case about the transfer of core value within the IT industry where the Internet service providers become predominant. In the past, the telecom giants (e.g. China Mobile, China Telecom and China Unicom) were formidable and unattainable. These telecom giants are massive in term of scale, value of fixed assets, business turnover and profits. However, a thorough analysis reveals that business and service functions of these telecom companies are actually relatively simple. They place focuses on providing just basic communication services such as phone calls, text messages and ram mobile data to the consumers of businesses, individuals and society. They are lack of innovations in technologies, service mode and novel approaches to achieve better values. Meanwhile, most of their equipment and technologies are purchased from other suppliers. As a result, these telecom giants are deficient in values and heavily rely on monopolizing resources. Indeed, their core values mainly derive from the network constructions and operations' permit authorized by the government. With relatively mediocre profits, they primarily live on the massive market demand and monopoly in China. As a result, their stock prices and market values are not as high as expected. Later, through constant technological innovations, the telecommunications equipment suppliers (e.g. Chinese Huawei and ZTE) have gained rapid growth in scale, sales, profits and share prices. With strong competitiveness, they have become renowned transnational enterprises at the global scale. However, it is worth noting that, even though Huawei and ZTE become the leading hardware manufacturers and suppliers in telecommunication industry, their market positioning is telecom equipment manufacturer and supplier. This has resulted in their lower-than-expected share prices and market values. It is necessary that they transform themselves from the "asset-heavy" enterprises to the "deindustrialization" and "asset-light" industrial Internet companies. In addition, they should shift to the service-centered development mode from their traditional operation mode (i.e. production and manufacturing oriented). To be frank, I really expect the revolutionary transformation of these enterprises in the future.

The Internet economy era has arrived. Tencent was once a tiny and inconspicuous company in China which was engaged in Internet games. In the blink of an eye, Tencent has achieved an astonishing market value of as high as 530 billion US dollars through innovations in technologies, service experiences and business mode.

Similarly, Tencent developed WeChat (similar to Facebook). As a result, Tencent has won full recognition from the global capital market despite its business revenue and scale are still far behind that of China Mobile, China Telecom, China Unicom, Huawei, and ZTE. Indeed, Tecent has accumulated a large amount of funds, and possessed a powerful strength and opportunity to dominate the future market. By analogy, China Mobile, China Telecom, China Unicom are more like aircraft-carriers or large ships sailing in the vast ocean; Huawei, ZTE are more like high-speed trains running on the rails. By contrast, Tencent's WeChat is more like an aircraft flying in the sky, which is faster, higher and further in distance. With extraordinary flexibility, this aircraft could soar freely in the boundless sky. Nearly everyone wants to board on such an aircraft to receive its warm and attentive services from stewardesses. Everyone is willing to accept all services provided by this aircraft. Although WeChat seems like a "tiny" App, its service has touched the heart of every Chinese people by connecting together your family members, relatives, classmates, friends, colleagues, business partners and other contacts in a timely and convenience manner. Friends are enabled to chat, make friendly contacts, talk, communicate, share information and news, and work together at any time in a pleasant atmosphere. Wechat evokes love and care in such cold, harsh and competitive world. Similarly, it brings warmth, happiness and comforts. By providing such heart-warming services and experiences, WeChat has already attracted billions of loyal users, which fully reflects the true value of Internet. Meanwhile, WeChat has received massive trading funds and deposit funds. As an Internet service company adopting "asset-light" industrial development mode, WeChat has attracted a huge amount of investments and has developed into a super-growth company. It has begun the process to expand its territories to smart city, smart planet, media industry, e-commerce, the real economy, software industry, mobile office, enterprise informationization and other industries. Thus, its future is immeasurable. A snake that can swallow an elephant. Similarly, it can be predicted that Tencent can even effortlessly acquire a certain telecom operator if it wishes. In other words, today's monolithic telecom operators and telecom equipment providers are in fact working for Tencent, an Internet company. Despite small in scale, Tencent is able to collect all profits and values in the Internet industry.

As shown in Table A.10 (Appendix A), the market value of Tencent's WeChat (an Internet service provider) has experienced an amazingly rapid growth. It has accumulated tremendous capital and wealth in such a short time.

WeChat means much more than telecom operators' text message. In fact, WeChat means the experiences of application and service, smart enterprise, smart city and smart earth. Internet economy era has arrived and the global economic development mode has changed. Meanwhile, the service industry and services have undergone significant transformation. The Internet companies will completely transform the traditional manufacturing industry, service industry and financial investment industry. Different kinds of services have different prospects. Indeed, software, Internet, big data and AI generate core values. These core values are flowing to the innovative services, to the WeChat, and to the software industry, big data, AI and Internet industry. In fact, "Internet+" is a new service mode and a new economic theory.

8 Different Kinds of Finances: Software, Internet, 5G, Big Data, AI and Block-Chain Will Reshape Financial Industry

In China's planned economy era, all banks and investment companies were state-owned. Under normal circumstances at that time, enterprises should raise their requests to the corresponding authorities if loans and investment support were required for their production and service activities. Upon the approval, financial institutions (e.g. the banks and investment companies) would make loans or investments available. Therefore, the business relationships between financial institutions and enterprises were fairly simple. They were independent of each other while everything was subject to the plans and decisions made by the state and levels of government authorities. There was neither service concept, nor business integration between financial institutions and enterprises.

In fact, such a "financial culture" under this kind of business relationship has been affecting China's banking and investment industries all the time. Since the reform and opening-up in 1978, despite a series of institutional reforms, today's business relationships between financial institutions (e.g. banks and investment companies) and enterprises are still dominated by purely financial lending and investment. The financial investment industry derives its earnings mainly from interest margin. In other words, banks and investment companies are generally not deeply involved in the business development of enterprises. This is an extremely simple and obsolete development mode for the financial businesses and services. What methods and means can be used by banks and investment companies to control their investment and lending risks? Borrowers do not receive any business supports from banks and investment companies except the funds. As a result, the project success depends solely on enterprises' own efforts. Therefore, there are significant investment and lending risks associated with the financial investment industry in China. This has regularly resulted in their business failures. Meanwhile, the chain debt is a common phenomenon which presents a significant barrier to the healthy development of financial investment industry in China.

There was another confusing issue in China's financial investment industry which also violated the laws of market economy. During the planned economy era, there were a large number of national or regional banks and investment firms that were horizontally oriented. Even today, a large amount of such regional banks and investment firms are still being approved and established by the governments. Generally speaking, these national or regional banks and investment firms have no sense of concepts for their business' operating in some specific industries or sectors. They may involve in any industries or sectors, offering loans and investments to any sorts of projects. So, what methods, means and tools can these seemingly powerful and magnificent financial institutions use to control their loans and investment risks? In fact, quite a considerable number of regional banks and investment firms even have become the administrative resources of local governments. They do not have access to any effective means or methods of risk control. Eventually, they developed into a structural issue in China's financial industry and economy. They impeded China's pace to the market economy and modernization, failing to provide strong supports to the rapid development of enterprises. To deal with this situation, we could draw lessons from the United States and other Western developed countries by meeting the needs of market economy development. These developed countries have made great efforts in establishing "vertical management" professional banks and investment firms that are industry or sector-oriented. These banks and investment companies are deeply involved in enterprises' business development. Consequently, these professional banks and investment firms have become the key forces to lead the industrial Internet that is oriented to those sectors or industries. To survive and sustain development, the Chinese banks and investment firms must be very familiar with and proficient in the industries, sectors and the business. In this way, they will be capable of providing loans or investment to the appropriate enterprises with advantageous resources and market competitiveness in the intense market competition, thus mitigating the financial risks.

Let's have a look at practices of the US financial investment industry. In terms of business forms, the US banks (represented by Citibank and JP Morgan Chase) and the investment companies (represented by Buffett and Soros) are constantly transforming themselves into the professional orientations. They have increasingly focused on providing loans and investments to their familiar industries and fields. In terms of business development, they have shifted their core business mode from selling banks' own products to serving the customers and their businesses. They attach great importance to understanding customers' actual needs and pay attention to the supports needed by customers in their business development. In recent two decades, these banks and investment firms have been directly or indirectly making heavy investment in the development of software, network, big data, and AI. A large proportion of these software is in fact directly linked to the operation, management and business of borrowers or investees. Apart from massive funds, they also provide customers with the necessary management tools and business development tools that are based on networks, software, big data and AI. For customers, apart from funds, they can also continuously receive in-depth consultation services from the financial institutions on their management and business development. This is essential to the success of enterprises and investment projects. For investors, such a practice can enable them to obtain data or the big data on real-time operation, management and business development of customers at any time. By analyzing such big data and AI, these financial institutions can fully understand the status of customers in terms of operation, management and business development. As a result, they could easily and effectively make reinvestment or withdrawal decisions, thus significantly mitigating associated risks. Therefore, we should learn from these successful practices of the financial investment industry in the United States and other Western developed countries. It is likely that their practices can effectively deal with our challenges as well.

Today, the United States is fully promoting the development of "asset-light" industrial Internet companies to lead the development of global Internet economy. Therefore, the financial investment industry needs to be deeply integrated with the manufacturing, services, software, and Internet industries. This presents significant challenges to the traditional financial investment industry. So, whose banks and investment companies have achieved deep business integration with the manufacturing and service industries; and have made initial investments in developing information technologies such as networks, 5G, software, big data, AI and block-chain, they are more likely to have a smoother and faster transformation towards the Internet business and the industrial Internet business. Consequently, they would be more adaptable to the new development mode of Internet economy. There is no doubt that a significant amount of US financial investors are now leaders in the industrial Internet industry and Internet industry. In other words, they have experienced a simple and smooth transition into the Internet industry and the Internet economy.

In Beijing, there is a Canadian-funded software development company that is oriented to enterprise and business management. In the past, to catch up with the pace of information technologies, this company used traditional marketing method to promote its software products. However, it's not easy at all to sell software products in the Chinese market. Due to the absence of business background, this company suffered losses year after year. An market investigation of US and Canada revealed the new development mode of the US financial industry which is heavily focused on developing software, Internet, big data and AI with the business. As a consequence, the boss of this company made a decision to transform its development mode without any hesitation. This company still concentrates on developing a variety of software for enterprise management and businesses. However, the development mode and practices are entirely different from the past. From the market perspective, the software company and its products seemed to have disappeared. However, to the downstream, this company has developed a huge amount of software related to the manufacturing, service, management and business of enterprises. To the upstream, this company has established profound cooperative relationships with those international and domestic investors. In fact, this company has transformed itself into a financial investment and consultancy service business that deeply relies on software, Internet, big data and AI. One condition of its investment is that the investees, either manufacturing enterprises or service firms, must use the software and information systems developed by this company. These software have greatly supported the business operations, management, service and business development mode of investees that are relatively weak in the informationization. In addition, this company can obtain the data related to the management, operation, and service of investees at any time. This allows the company and its behind investors to control the investment risks by analyzing related data. This company's successful transformation in development mode has benefited both investors and investees. As a result, multi-win is achieved. Its strong support has been provided to the rapid development and expansion of Shuanghui Group, Little Sheep Hot Pot and other renowned companies in China.

There are so many excellent software and Internet companies in China such as Yonyou, Inspur and Kingdee, Alibaba, Tencent, MI, Didi and Baidu. In addition, there are a number of traditional banks and investment companies that desperately need the support of Internet, software, big data and AI. Therefore, it is imperative that Chinese financial institutions (e.g. banks and investment companies) deeply collaborate with those software and Internet companies so that a win-win outcome can be achieved.

Let's compare Chinese traditional banks and investment companies with Alibaba's Alipay and Tencent's WeChat Pay in terms of efficiency, cost, investment risk, service performance, analytic performance of big data, capital potential and popularity among youngsters. Internet economy is irresistible and is transforming the traditional financial investment industry. This is an urgent call for the in-depth fusion of Chinese financial investment industry with the manufacturing industry, service industry, software industry, and Internet industry. It is well recognized that "asset-light" industrial Internet mode is the future of development mode. The traditional financial industry is undergoing profound transformation. The information technology such as software, Internet, 5G, big data, AI and block-chain could create core values, which has become new methodology of financial development. In the past, traditional banks and investment companies were passive in making loans or investments, simply waiting for customers to apply for loans and investments. By contrast, today's Internet service "banks" tend to provide active services and in-depth business supports. They also actively explore the potential associated values of customers by providing a variety of Apps services in order to achieve the multi-win outcomes to the largest degree. It is now or never. China's traditional financial investment industry must move and act in accordance with the circumstances and times. Otherwise, it will lose the competitiveness advantage. The software, Internet, 5G, big data, AI and block-chain will definitely reshape and restructure the financial investment industry. The future of China's financial industry belongs to the Internet financial service industry.

9 Multidimensional Integration and Innovation Could Create the Core Values of Internet Economy

Traditional products and services are relatively monotonous and independent in forms, functions and values. For example, chairs and sofas allow you to sit comfortably; buses, trains, ships, planes can help you to travel; TV set plays sound & images for you; fridges keep your food fresh; washing machines wash clothes for you; air-conditioners adjust temperature for you; machine tools are designed to do machining work; mobile phone's original function lies in communication; computer's original function lies in computation; libraries allow you to gain access to information and data; the stadium provides a venue for sports and exercises; stores sell living necessities; manufacturing enterprises produce certain commodities or products for the society or other enterprises. These products, services, and activities are explicit in their functions, experiences and values. In fact, we are extremely familiar with the manufacturing model, service mode and value orientation of these products.

However, we are now in new era supported by software, Internet, 5G, big data and AI. In this new era, the relationships among aforementioned products, services, and activities have gradually shifted from a relatively independent "loose coupling" to an interrelated "tight coupling" in terms of forms, functions and values. In addition, they are more clear, compound and powerful in functions; and more intimate, warm and personalized in service experience. Their values are being constantly explored by means of innovations. Greater and broader values are being created, thus maximizing the values. Meanwhile, unprecedentedly tremendous changes have taken place in the structures and forms of enterprise, industry and economy, as well as traditional manufacturing, service and economic development modes. Similarly, industrial and economic values have undergone significant changes. In simple terms, a new economic era, i.e. the Internet economy era, has arrived. So, what are precisely the core values of Internet economy?

First, people and popularity are the ultimate values of Internet economy. To Service the people is the core essence of Internet spirit. So, those, who could perform innovations on technologies, products and industrial development mode; who could constantly providing valuable and innovative Internet services to make consumers be satisfied; who is capable of gathering people and popularity to the largest degree, and making everyone participate in innovative activities to create and realize values, would take the preemptive opportunity and be more likely to convince investors to inject a huge amount of money continuously; could stand on the commanding height of Internet and the Internet economy; would have the potential to change the world and to reshape the industrial landscape; would finally become the dominant leaders of the economy and the cornerstone of the nation.

Starting from search engines, Google and Baidu provided digital library services at the initial stage. Then they began to provide services related to smart city and smart earth such as navigation, machine translation, interactive community, entertainment, mobile services, as well as Apps and software development tools. In addition, they have started to explore the real economy, e.g. developing digitized, networking and intelligent-based automobiles as well as providing financial services. Similarly, Apple and MI provide products such as smart-phones, PADs, tablet computers, watches, entertainment terminal products and a wide range of Apps services and experience, thus having established huge industrial ecosystems. Nowadays, Apple and MI are also advancing into the real economy, e.g. developing automobiles, medical robots and digital home appliance products based on digitization, networking and intelligentization as well as providing financial services. Starting from e-commerce, Amazon, Alibaba and JD.Com have now entered the service industry to provide logistics, media and online medical services. In addition, they have entered the real economy to produce intelligent cars and provide financial services. By providing powerful services such as entertainment, WeChat, interactive community and App software, Tencent has connected groups of people, friends and classmates closely together through "tight coupling" mode. It has changed the ways of communication and interaction between human beings. Furthermore, Tencent is advancing towards the media, e-commerce, office software, enterprise information services, financial services and other domains. All roads lead to Rome. In fact, the objective of these

Internet companies is to compete for users and popularity. These companies bring together large crowds and perform classification, mining, and analysis based on big data. In turn, these companies can provide more accurate, appropriate, meticulous, and better services to more people. Meanwhile, they are constantly expanding their industrial ecosystems, eco-chains and innovation circles to the largest degree. Consequently, they can obtain maximum investments, highest profits, highest share values and highest market values. At the same time, the core values can be realized to the largest degree through deeply exploring the human needs.

Second, Apple company represented the "deindustrialization" and "asset-light" industrial Internet development mode, which places focuses on managing the core values of industrial ecosystem. This is entirely different from the traditional manufacturing industry's "asset-heavy" development mode that aims at managing the enterprises themselves and products. Thus, the industry's core values will be maximized whereas the operational risk of enterprises and the investment risks of investors can be minimized. As a result, Google, Boeing, GE, Siemens, MI and other companies are in a desperate rush to imitate Apple's new industrial development mode. Western entrepreneurs, industrial circles, financial capital circles and economic circles are all focused on studying Apple's success and its new mode. Steve Jobs has become a legendary in economy. Under the "Deindustrialization" and "asset-light" mode, industrial core values will be maximized through the innovations in the industrial development mode. Indeed, we are in an era of Internet economy which is featured with a much more advanced development mode. It can be concluded that such a new industrial development mode represents the core value of Internet economy.

More importantly, the software, Internet, 5G, big data and AI can not only improve the productivity of all industries and circles, but also change the production relations and even our life style. This will lead to the next industrial revolution at all levels and circles. Such a new the production relations also represents the core value of Internet economy.

Third, the traditional manufacturing industry has taken advantage of the control of quality manufacturing resources and core technologies of the manufacturing industry. By adopting the "+Internet" mode, the traditional manufacturing industry is advancing towards the industrial Internet, pursuing in-depth integration with software, Internet, 5G, big data, AI and chip industries. The transformation and reforms of the traditional manufacturing industry is achieved. Similarly, the Internet and software industries have taken advantage of the control of powerful information technologies (e.g. networks, 5G, software, big data and AI) and their super financing capacity. By adopting the "Internet+" mode, the Internet and software industries are advancing towards the industrial Internet. They are revolutionizing those traditional manufacturing industries through the in-depth integration with the manufacturing and associated services. The "+Internet" mode and "Internet+" mode will interact with each other in the industrial Internet industry. In summary, they are different ways leading to the same destination. In the future, the manufacturing industry and its associated service industries will be deeply integrated with the financial investment, Internet, software and chip industries. Meanwhile, all the business will be closely integrated with Internet, 5G, software, big data, AI and other information technologies. Also, the technologies will be closely integrated with "asset-light" industrial development mode. These all will lead to develop the industrial Internet industry which is a new multi-industry-dimensional integrated form. So, the multi-dimensional integration and innovations present the core value of Internet economy.

Fourth, those who control the money, capital and finance are able to control enterprises, industries, economy, and even the world. Today, those Internet companies have gathered vast amounts of users, popularity, transactions, deposit funds, and a huge number of youngsters. This is achieved through software, networks, big data, AI, the business and technological innovations, product innovations, industrial development mode innovations, the human-oriented and personalized services, and the deep exploration of human needs. The Internet-based online money transactions are proceeded at lightening speed, hundreds of times faster than that of the traditional banking and financial industry. At the same time, the Internet companies have received a great deal of investment opportunities from the financial capital industry and the stock market. Through controlling those massive money and investments, the Internet companies are transforming themselves into new leaders in emerging financial investment industry, which has changed the development mode and dominant position of the traditional financial industry. This will finally change and subvert the financial industry's form that has been built for more than a century. This is also the core value of Internet economy.

The survival of traditional financial industry depends on its close cooperation and integration with the Internet industry and industrial Internet industry. Today, the financial capital has accelerated its flow from the traditional financial capital industry to the Internet financial service industry, which is irreversible.

It is worth noting that the economy should be dominated by capital and finance, rather than the manufacturing industry. Those with more capital are likely to be given more opportunities in future economy and industries, thus obtaining the competitive edge. Therefore, the industrial and economic values will tend to flow towards those industries with high profits and share values. In the era of Internet economy, a large amount of capitals and values are increasingly flowing to high-end manufacturing industry, chip industry, software industry, Internet industry, and industrial Internet industry. As a matter of fact, only innovation can generate values.

Value has different meanings in different eras. Endless innovations and values will be generated through the deep integration of software, Internet, 5G, big data, AI and manufacturing industry, financial investment industry and service industry as well as the new industrial development modes. As a result, the format and structure of traditional businesses, industries and economy will undergo radical changes, so as the traditional value system of industries and economy.

To those who engaged in the traditional manufacturing industry, they regard the manufacturing industry as their own realm and cause. They naturally would subsist on and earn money from the manufacturing industry. However, today's financial industry and Internet industry have closely integrated with together. Due to the profit-driven nature, they will raise large amounts of capitals and advance into some real economic sectors with potential benefits. From their point of view, the manufacturing industry and the real economy are simply means and carriers for making money and realizing capital efficiency. They capitalize on the industrial Internet mode to realize profits and values. In general, the financial industry and Internet industry are not directly perform manufacturing activities. Rather, they engage the manufacturing industry to perform manufacturing, processing and assembling for them which involves a tiny amount of profits. Moreover, they are good at collecting vast majority of the profits as well as all share values and market values of the industry. This is achieved through the deployment of capital and the allocation of global advantageous manufacturing resources. Not matter you like it or not, it is becoming a reality. This is the essence of "deindustrialization" and such new industrial revolution.

Similar to the animal world, law of the jungle also applies to the human society. Similarly, law of the jungle applies to the relationships between countries, industries and enterprises. All of them have to secure its survival and development by relying on cooperation and competition. Therefore, rather than sitting still, the traditional manufacturing enterprises should wake up and adopt the "+Internet" mode. Through making strenuous efforts and transforming themselves into the industrial Internet businesses, manufacturing enterprises can keep pace with the development of Internet economy and control their own fate. It is the time for China's monolithic traditional manufacturing, service and financial industries to awaken.

Chapter 7
Critical Moment for China's Economy and Industries

The manufacturing industry, service industry, financial investment industry, Internet industry and software industry are integrated with each other closely in the United States, Germany and other western developed countries. Their leading enterprises are transforming themselves towards the "asset-light" industrial Internet development mode. By contrast, China's industries and enterprises are still dominated by the "asset-heavy" growth mode. Constrained by UN's Paris Agreement on Emission Reduction, China's industries and enterprises are under enormous pressure. They will have to make a choice that determines their fate.

1 China's Industries Have to Make a Choice that Determines Their Fate Through Transformation

During the more than four decades of reform and opening-up, China has achieved remarkable achievements in economic development. China's GDP has surpassed $10 trillion, ranking the second in the world. With a prosperous and thriving market, the output value and scale of most industries maintain the first place steadily at the global scale. The last decades have witnessed a surge in national strength, rapid development of enterprises and continuous improvement of competitiveness. Indeed, people's demands are constantly fulfilled.

The achievements are hard won. A revisit of our achievements from a global perspective reveals that these results have mainly derived from the Chinese manufacturing, processing industries and associated service industries that are dominated by traditional "asset-heavy" growth mode. These achievements are made from the spirit, the tireless efforts and hard work of the Chinese people. However, we have to admit that the growth is on the high cost of resources, environment and manpower.

Since the admission to the World Trade Organization, China has become a "good student" with rapid economic development and an increasing share in the international market. However, we have to admit that China's major industries are still

J. Yuchi, *Value Proposition*,
https://doi.org/10.1007/978-981-15-5025-6_7

concentrating on those low and medium-end manufacturing, processing and service sectors. Most Chinese enterprises are engaged in low value-added works that the leading western companies are no longer willing to do. They gain profits from low value-added processing costs, labor costs and service charges. Compared with enterprises in western developed countries, most of Chinese enterprises are generally inefficient and unprofitable while mainly relying on large domestic demands. A typical example is Apple's industrial ecosystem and labor division. Apple and its core chip makers reap most of the profits, leaving around 4.6% of the profits for Chinese manufacturing and processing enterprises. In this process, we did secure jobs and wages for workers; we did enjoy a rapid growth of output value and GDP. However, our industrial and corporate profits are extremely low. China, a country with a population of nearly 1.4 billion, is still at relatively early stage of the market economy. It is now even difficult for us to secure those outsourced processing works, which can at least generate a small fraction of profits and money. Today, relatively developed regions in China (e.g. Guangdong, Zhejiang, Jiangsu and Shandong provinces) have become one of the world's largest "processing factories" and "processing zones", and one of the major links of international supply chain. By contrast, China's vast western regions are still struggling to get a share of the "feast". In such process, a great number of Chinese enterprises have invested a huge amount of precious natural resources Similarly, the state and society have paid enormous environmental costs, including water and air quality. These natural resources are indispensable for the survival and development of human beings. It is fair to state that this is the current positioning of China's enterprises, industries and economy in the broad economic globalization set by the United States, other western developed countries and their leading companies.

If China's economy and industries are always satisfied with the status quo, they can simply accept the positioning set up by the western countries and their leading companies in the global industrial pattern, i.e. low-to-medium-end processing, manufacturing and service industry. As a result, China's economy and industries will always be controlled by with limited development opportunities. In the end, it would be difficult for China's economy and industries to achieve a sustainable development. In addition, the state and enterprises will remain "big but not strong", continuing to be exploited by the Western economic powers. Therefore, China's economy and backbone enterprises have to completely transform themselves to achieve a "rebirth". This is the historic mission for every single Chinese entrepreneurs, innovators, and economists. Our businesses, industries and economy have to make a choice that determine their fate.

2 An Enormous Economic Situation Pressure on China

From a macro perspective, due to the lack of growth momentum and demands, the US, Japanese and European economies will remain in recession for a long period of time. This tendency and situation will not be changed in short term. This means that, to expand and encourage exports which is one of the major drivers of the economic growth for China, will be faced a long-term pressure. At the same time, there are always significant difficulties in exploring domestic demands and encouraging domestic consumption. As a consequence, there is enormous pressure in expanding China's domestic market, which is another major driver of the economic growth. Meanwhile, it will take a long period of time to transform those Chinese enterprises and to optimize China's industrial structure. Therefore, it will be a long-term keynote that China's economic development will continue to rely on investment, manufacturing industry and export processing industry. However, if this process takes too long, the quality of China's economic development and the enterprises' sustainable development will be inevitably affected.

In fact, it is also not that easy for the Chinese manufacturing industry to secure the position of low-to-medium-end processing, assembling and service industries at the global scale. With rising costs (e.g. land and labor), China is constantly losing dominance in manufacturing, processing and assembly industries in the international arena, which is continuously transferring to the developing countries in Southeast Asia and South Asia. This warrants alert and attention. There is pressure from everywhere. As a result, it is really difficult to secure such a "cake" even with a small profit.

From the medium and long-term perspective, China must cope with the pressure from United States and other western developed countries to honor our commitments to United Nations Framework Convention on Climate Change. At the moment, China's economy is still dominated by the manufacturing, processing, assembling industries and low-value service industry that is featured with "asset-heavy" growth mode. Therefore, it presents a significant challenge to fulfill our commitments in terms of energy conservation and emission reduction.

From the perspectives of in-depth level and internal causes, most of Chinese enterprises have not yet completed the industrialization, i.e. the most fundamental process of economic development, not to mention the critical process of informationization. Meanwhile, China lags much behind in the high-end manufacturing. Those high value-added products are rare and heavily dependent on foreign imports. In addition, vast majority of the profits have flown to the western developed countries such as the United States, Germany and their leading enterprises. Today, driven by the Internet economy, the global economy is undergoing profound changes. Further, only very few Chinese enterprises are taking the new industrial Internet development mode, which draws more concerns. To sustain development, China's economy and industries must have core competitiveness to participate in the international competition. To win the respect of those international counterparts, we must identify those internal factors that drive the development, and develop technologies, market, industries and economy by relying mainly on ourselves.

Due to the accumulation of aforementioned disadvantageous factors, we can clearly feel the mounting pressures derived from the economic slowdown and downturn. Today, the pressure has spread to China's economic lifelines, i.e. the capital market and the stock market. The capital community, in particular the international capital community, does not recognize the huge number of listed "asset-heavy" enterprises that are generally with low efficiency. Those capital communities are not willing to make long-term investment, which is reflected in short-selling of the shares of listed enterprises in China. This has resulted in drastic fluctuations of the Chinese stock market. Sometimes the Chinese government has to "intervene" during stock crises. As a result, the economic and financial stability, people's livelihood and social stability in China are severely affected.

3 Under Enormous Pressure From Western "Hegemons", We Must Resort to Restructuring, Transformation and Upgrading

It is well recognized that the Chinese economy is presented significant challenges which requires the adjustment of economic structure as well as the transformations and upgrading of enterprises. Meanwhile, the Industry 4.0 (Germany) and Internet and Industrial Internet (US) have become a heated topic in industry, capital and economic communities in China and beyond. These two events seem to be uncorrelated. However, a careful analysis reveals that they are in fact two sides of a same coin. They represent the different stages of growth and development of industries and businesses. The serious discussion and analysis have been carried out in earlier chapters of this book. Hopefully this provides some inspirations and references for the development of China's economy, industry and business.

The United States, Germany and other western developed countries have been leading the global economic and industrial patterns. Under enormous pressure from those western "Hegemons", we must resort to a great transformation and upgrading. We have previously examined the objective and essence of Germany's Industry 4.0; the fundamental of US third industrial revolution as well as the emergence and development of "asset-light" industrial Internet development mode. Since China didn't start the development of market economy until the reform and opening-up in 1978, we should recognize that China's economy and industries must resort to a smart strategy and wisdom. As a result, we can relieve the intense pressures from the United States, Germany and other western developed countries; and transform the economy from the "big" to the "strong". Therefore, on the one hand, China's industries and enterprises need to find the root causes of economic structural problems and the appropriate methodology for economic restructuring with a consideration of the uniqueness of Chinese economy. In addition, efforts are required to identify the appropriate pathway for the transformation, upgrading and elevation of industries and businesses. Furthermore, the new economic and industrial development mode must be explored. China's future industrial and business formats must be properly perceived and controlled. Meanwhile, China's future economic theories must be clarified. As a consequence, an explicit description of China's future economic structure and blueprint can be made. On the other hand, we need to learn from western developed countries such as the United States, Germany, and their leading enterprises with an open mind. Moreover, we need to seriously analyze their economic and industry development history, laws and modes. In addition, we need to cooperate with western economic powers to study, discuss and practice the methodologies to achieve economic prosperity. Ultimately, China must find a suitable development pathway for the economy, industries and enterprises.

Thus, despite enormous pressure from those fierce competitors, China's economy, industries and enterprises must be adaptive and active in the restructuring, transformation and upgrading. In a word, China's economy today must deal with the challenges and opportunities derived from the era of Internet economy.

Chapter 8
Internet, Big Data, AI and E^3 Economic Theory Are Driving Forces to the Transformation of the China's Economy

What are the development processes of economy, manufacturing and financial industries in China? In essence, what are the structural problems of China's economy? How have these structural problems influenced China's principal economic sectors (i.e. manufacturing and financial investment industries) and the stock market? What is the significance of informationization on China's development of the Internet economy? What is the rationale behind the statement that the Internet, 5G, big data, AI and "e^3" economic theory are the driving forces that can change China's economy?

Recently, China's economic restructuring has attracted attention in the newspaper or on the Internet. It was stated that the proportion of China's economic structure is gradually leaning from the manufacturing industry towards the service industry. There is no doubt about the broad direction. However, the key lies in the manufacturing industry rather than the service industry. Can we resolve the economic restructuring issues in such a simple and fast manner simply by significantly increasing the proportion of service industry and decreasing that of manufacturing industry? Can we easily deal with the "middle-income trap" simply by doing this? Is it really that simple? Indeed, we will be fooled by some so-called western experts if we really believe it.

In China, manufacturing industry is the backbone of national economy. In my opinion, the industrial development mode and production relations of China's manufacturing industry must meet the new requirements of Internet economy in terms of development mode and theory. This is the key of adjusting China's economic structure. In other words, Chinese manufacturing enterprises should shift their focus from the product-orientated to the service-orientated. Meanwhile, they should focus on improving efficiency, reducing costs and significantly enhancing competitiveness. Most importantly, the products must feature with high quality, low prices and excellent services. Only this way can we survive in the intense market competitions by means of meeting the needs and demands of the market. Consequently, the services industry should place more focuses on selling our own products, instead of that of others. The structural problems of China's economy cannot be truly solved until the relationship between the manufacturing industry and service industry is optimized.

J. Yuchi, *Value Proposition*,
https://doi.org/10.1007/978-981-15-5025-6_8

In fact, the predominance of economies, industries and enterprises in the United States, Germany and other western developed countries lies in the powerful root of their real economy. Indeed, their manufacturing industries have controlled the core technologies and standards of the industry. They constantly realized the industry's core values and adopted the new industrial development modes through continuous innovation on production relations. They seem to have no or a small fraction of manufacturing industries or sectors. In fact, occupying the high-end of the industrial value chain, their manufacturing industries are tightly controlling the global manufacturing, associated service and financial investment markets. Therefore, the key of economic and industrial structure adjustment in China does not simply lie in raising the proportion of service industry in the national economy. Rather, it lies in the innovating and controlling the core technologies and standards of the manufacturing industry based on the strong support of integration with information technologies (e.g. the Internet, 5G, software, big data and AI). Similarly, it lies in the thorough transformation of the traditional industrial development mode to the new industrial Internet development mode. Once the manufacturing industry is properly adjusted, its associated industries (e.g. service and financial investment) and the stock markets will naturally be back on track. Therefore, the manufacturing industry will always be the key.

1 What a Hard Journey China's Manufacturing and Economy Have Experienced?

The first period was from the feudal economy to the early stage of capitalist economy. At the beginning of the 20th century, with the cease of Qing dynasty, China's feudal society came to the end. In other words, China didn't start the transition process from feudal economy to so-called capitalist economy until one century ago. Over the next four decades until the founding of New China in 1949, China had experienced incessant wars between warlords, Anti-Japanese War, and the war of liberation. As a result, the nation was devastated; people were in dire poverty and on the edge of starvation; and economy collapsed, let alone the existence of any industry or industrial system.

The period from 1949 to 1978 marked the development of the planned economy and industrialization. When New China was founded in 1949, many things needed to be done and many industries needed to be revitalized. However, the nation had to deal with the Korean War and long-term economic blockade by the US and western countries. As a result, significant challenges were presented to the national development and the people's living demands. Later, since the First Five-Year Strategic Plan, the planned economic system of the Soviet Union has been adopted as the economic development mode in China. Under the guidance and assistance of the Soviet Union, a large number of large state-owned enterprises and collective SMEs were directly founded and set up by the state and local governments to engage in heavy industry, light industry, commerce and banking business. It is a valid statement that the planned economy had played a crucial role in addressing the development needs of all industries and trades and in easing the pressure from people's living needs.

During this period, we also experienced the "cultural revolution", while the college entrance examination was suspended. In fact the economy was on the verge of collapse. In addition, China's foreign exchange reserve in 1976 was so low that it was only enough to purchase several Boeing aircraft.

The miserable modern history and economic development history of China are briefly analyzed. My intention is to tell everyone that China's modern industrial history and economic history didn't start until 1949, when the planned economic system of the Soviet Union was adopted. Under the internal and external environment of that time, the planned economy substantially satisfied China's economic development needs and people's living demands, and helped China to build a preliminary industrial system and foundation. It is a valid statement that the implementation of such a planned economic system was the only choice for China's economy in the broad context of international politics and economy of that time. The advantage of planned economy was that the state could pool resources to solve major problems, addressing the nation's critical needs and the basic needs of the people within a short time. However, the drawback lay in the low efficiency of industry and enterprise as well as the high social costs. Compared with the market economy of western developed countries, it would be easier for China's economy, industries, and enterprises to grow in size than in competitiveness. Therefore, it is inevitable that conflicts exist between the planned economic system implemented by China in last few decades and the mechanisms of market economy with unique characteristics of China today. Indeed, China has to continuously make adjustments through constant economic reforms.

During that period, due to the long-term economic blockade imposed by western countries, it was extremely difficult to manufacture major products needed by the state. Similarly, it was enormously hard to supply living necessities to the people. Therefore, it would inevitably lead to the lack of competition awareness, cost concepts and service awareness. This has resulted in weak management of manufacturing enterprises as well as poor services of commercial enterprises and banks. It was difficult to form a normal business relationship between the manufacturing enterprises and the banks. Everything depended on the nation's administrative command. Enterprises were large and all inclusive. Like a society, enterprises were all-embracing. The state in fact managed everything and the whole process related to the production activity of each enterprise, no matter its size, from beginning to end. In addition, there was no concept of product quality, price and services. There was no concept of entrepreneurs, who actually were government-appointed. Under the planned economy system, some considerable progresses were made in economy, industry and business, while China's own industrial system was preliminarily established. However, China's economy didn't present features of modern market economy, such as socialized division of labor, specialized development and sufficient competition. The economic concepts were too simple. Therefore, the emergence of structural problems of China's economy is bound to have its historical reasons.

Since 1978, China's economy has been gradually advancing towards the socialist market economy with Chinese characteristics. In 1978, China began the reform and opening-up to the world, ushering in the rapid development of economy and science. The contract system was implemented in rural areas. In such new system, the development of township enterprises and collective economy were encouraged; privately or individually-owned businesses were allowed, and the pilot projects of private enterprises and private economy were carried out in Zhejiang province. Since then, a large number of successful entrepreneurs have emerged. Nowadays, some of them have become the pillars of national economy. In 1977, China restored the college entrance examination. The light industry and textile industry were relatively easier to develop. Through the development of light industry and textile industry, China solved the employment problem for the large amount of labor forces and population. This has increased foreign exchange earnings through export expansion, and also satisfied people's demands, which finally led to revitalize and reform the industrial systems for textile and light industries. In the 1980s, China began to learn the experience of advanced technologies and economic development from the western developed countries. Through mobilizing resources of the entire nation, China introduced Japan's steel production lines and technologies, and consequently created the modernized Shanghai Baoshan Iron and Steel Co., Ltd. Since then, China has begun to more open to the world. In the late 1980s, some household appliance production lines were imported from Japan, Germany and other developed countries to meet the growing demands. At that time, the United States, Germany, Japan and other developed countries, as well as their enterprises and financial industry provided a great deal of assistance and support. Subsequently, a large number of export processing zones, economic and technological development zones emerged, followed by a large amount of joint ventures as well as cooperative and wholly foreign-owned enterprises.

Later on, along with the intensive market competition, a fraction of Chinese enterprises have embarked on the right track of market economy. Quite a few of them have begun to undertake outsourced manufacturing, processing and service by joining in the global industrial supply chain and becoming a key link. A large number of backbone enterprises (e.g. Haier, Midea, Huawei, ZTE, Sany, Zoomlion, Lenovo, Inspur, CRRC, CSIC and AVIC) have gradually grown into large multinational enterprises through constant innovations. Similarly, supported by the national policies on encouraging innovations, a number of high and new technology development zones were established, where the high-tech enterprises were created and developed. It is a valid statement that since the early 1990s, China's manufacturing industry has been gradually advancing towards of market economy, accompanied by the increasingly healthier development and stronger market competitiveness. As a consequence, China has become a top manufacturing nation in terms of scale and output value. China's GDP has exceeded $10 trillion, and ranked the second, only behind the United States.

Since the founding of New China in 1949, especially over more than four decades since the reform and opening-up from 1978, China's economy and manufacturing industry have made remarkable achievements. China has rapidly completed the industrialization process within 60 years, in which those western developed countries have spent about three centuries. In fact, the implementation of planned economy has made a remarkable contribution. Within merely four decades since the reform and opening in 1978, China has gradually embarked on the market economy which has been taken more than 300 years in the western developed countries. However, it is worth noting that China's transition to the market economy went through very short period of time, which was subject to the long-term impact of planned economy. Due to such impacts, a considerable number of factors related to China's the economic, industrial and enterprise development are contradictory with the general laws of market economy. Indeed, these factors have presented significant challenges to China's economic development which need to be adjusted through constant economic reforms.

Looking into the future, China's economy, industries and enterprises need to gain strength in capacity instead of in size, meeting the high demands of participating in the broad international market competition. They have to adapt to structural changes in the international economy and industries. They have to take into account the fact that the United States, Germany and other western economic powers as well as their leading enterprises have begun their rapid transformation into new "asset-light" development mode of Internet economy. Therefore, it is imperative to have a profound discussion and understanding of the structural problems and contradictions that still exist in current Chinese economy and industries. In addition, we need to accelerate the reforms as well as solving these problems in order to meet new challenges posed by the global Internet economy.

2 Root Causes of Structural Problems in China's Economy

Despite several reforms and adjustments, most of Chinese state-owned enterprises (hereinafter as SOE) are still large-sized and all-inclusive in structure, tending to adopt the internal-cycle and "asset-heavy" industrial development mode. Compared with western companies, the Chinese SOEs are less competitive, big but not strong. Careful analysis reveals some clues. Let's carry out a general analysis of some problems and negative phenomena that often occur in China's SOEs. It should be noted that not every business has so many problems. At the national level, policies and measures have been continuously introduced to reform and adjust these issues. These enterprises are expected to make structural adjustments to transform and upgrade themselves as soon as possible.

Large and Monopolistic. At present, some of Chinese industries are open to private enterprises and the society at a low level. A considerable number of SOEs mainly live on monopolizing resources, being nurtured in incubators for a long time by the state. These enterprises are lack of the sense of competition in the market economy. In today's global economy, these enterprises have to improve their competitiveness, undertake the responsibilities as the mainstay of market economy, and participate in the international competition.

Monolithic and Chaotic. The SOEs were originated from the planned economy and were created by the state. As a result, a considerable number of entrepreneurs came from the government agencies. In other words, those entrepreneurs and government officials are extremely familiar with each other, who also have frequent job rotations within SOEs. Therefore, the inherent thinking could easily form between the SOEs and the governments, and between SOEs. This has resulted in the lack of "fresh blood" and advanced management experience, which consequentially makes them be reluctant to carry out the structural reforms.

At present, the SOEs' owners still have a vague understanding of self-consciousness. For example, they even don't clearly know who represent the state, who are shareholders, and to whom they should report. The responsibilities for the parties exercising true ownership rights are not fully in place, even missing. In this management mode, the huge amount of national assets and resources are managed only by several managers, relying solely on their self-regulation and conscience. Due to the lack of the mechanisms for monitoring and protection, these enterprises are extremely vulnerable to the corruption, resulting in loss to the state. Over time, if the event of economic growth, all contradictions would be concealed and everything would seem quite normal. On the contrary, in the event of economy downturn and depression, all the contradictions and problems will be exposed. Meanwhile, if SOEs' owners and operators failed to achieve the economic annual target or suffered a loss, they can simply blame the macroeconomic situation as an excuse. This is a common phenomenon. Indeed, these problems and vague understandings have to be fixed as soon as possible.

In terms of business operation and management, multiple management levels have resulted in low efficiency and high cost of the operations. The SOEs are lack of the sense of responsibility to shareholders while the procedures and responsibilities of their decisions are not clear. In private enterprises, the relationship between workers, managers and employers is bound by contracts. Entries and exits of personnel are implemented according to the labor contract laws. Compared to private enterprises, The SOEs' workers and managers often perceive themselves as employees of the state. In fact, it is very difficult for the operators to lay off any workers or managers. All managers are classified by ranks that are recognized by the state. Such a phenomenon has posed a significant challenge to the operators of SOEs, which has resulted in high business management costs.

The SOEs' operators and managers have a specific tenure. As a result, they are bound to focus on short-term results and featured with short-term behaviors. There is lack of long-term planning and consideration. Consequently, the SOEs attach no importance to the innovations and long-term behaviors. In the long run, they will be short in core competitiveness. In 2014, as a leading private enterprise in China, Huawei's R&D investment was up to $ 6.5 billion, accounting for 14.2% of sales revenue of the year. It is even greater than the sum of R&D investment of nearly 400 A-share firms.

Large and All-inclusive. A large number of SOEs have most of their production activities completed within their own corporation by adopting inner cycles, no matter the scale or the level of difficulties. This has led to excessive investment of SOEs in production equipment, facilities, industrial lands and staffs, forming "asset-heavy". These enterprises seem to be in spectacular growth momentum. However, in reality, they will suffer from high asset-debt ratio, high equipment depreciation rate and heavy personnel burden. Failing to allocate the national and global advantageous resources, these enterprises are lack of competitiveness. As a result, these SOEs have to bear all their own risks, not being able to transfer their risks to the society and the world. Such an industrial development mode has resulted in high business costs. This works well in the event of economic growth. However, in the event of economic downturn, these SOEs will suffer as a large amount of equipment, facilities, plants and staffs remain idle which cannot be used by the society. These SOEs will likely experience heavy debts and significant challenges.

Large and Complex. A large number of SOEs are overstretched, engaging themselves in all industries and trades. As a result, there is lack of concentration, professionalism and refinement. Consequently, these enterprises did not concentrate their precious financial resources on the core technologies, core values and high-end manufacturing of the industry. Over time, the SOEs' advantages will be weakened, which makes them difficult to maintain the market of high-end manufacturing.

Large and Disordered. Some SOEs are heavily reliant on non-core business, such as land expansion and real estate business development on a long-term basis. As their non-core businesses occupy a higher proportion, these SOEs' primary business will be indirectly affected in the event of fluctuations of the capital market or land policies.

A case study will be undertaken to analyze the typical dilemma faced by the state owned enterprises that are located in the northeastern region of China. Tables A.11 and A.12 in Appendix A show the GDP growth rate of three northeastern provinces in 2014 and the first three quarters of 2015. It can be observed that the GDP growth rate of three northeastern provinces ranked well below the national average. This is a worrying sign.

Ordinarily, compared to other provinces (e.g. Guangdong, Fujian, Zhejiang and Shandong), those three northeastern provinces should have advantages in infrastructure, facilities and conditions. These include the industrial foundation, transportation, energy, ports, geographic locations, human resources, etc. The key problem lies in the "large and all-inclusive" industrial structure of three northeastern provinces as well as the transformation and upgrading of related state-owned manufacturing enterprises. The economy and industrial structure of Northeast China have been centered around and reliant on the large state-owned manufacturing enterprises for a long period of time. As a result, whenever the output value and product sales of Changchun FAW's automobile slide, the economy of Jilin Province will be adversely affected. Whenever the output value and product sales of Daqing Oilfield and Harbin Electric Corporation decline, the economy of Heilongjiang province will be negatively influenced. Whenever the output value and product sales of electromechanical and manufacturing enterprises drop, Liaoning Province's economy will go wrong. A careful analysis reveals that the three northeastern provinces heavily rely on those large manufacturing enterprises, most of which are the "asset-heavy" and state-owned enterprises. This has resulted in a strong traditional manufacturing culture and a number of so-called "courtyard-style" cultures. Although a series of reforms have been implemented, these enterprises have been taking traditional industrial development mode that is featured with large and all-inclusive, inner-cycle, low-quality-and-low-efficient and "asset-heavy". In the event of economic growth, due to the massive Chinese market demands and product sales, those problems and contradictions are overshadowed. However, when the economy goes to downturn, problems associated with the systems, mechanisms and industrial development mode of enterprises will repeatedly emerge. In addition, the contradictions will become severer which presented significant challenges to the society and all levels of government. As a result, the economies, livelihoods and employments of the three northeastern provinces are affected. Due to the recent tough macro environment, the economic growth has slowed down which is associated with growing pressures from the economic downturn. Compared to other provinces in China, the economic performance of the three northeastern provinces is quite low. We can imagine that all levels of government of the three northeastern provinces as well as those large manufacturing enterprises are expecting the better macro economic environment so that they can survive and get through this difficult period of time. However, even if the macro economic environment may come better in the near future, will the economy and manufacturing enterprises in the three northeastern provinces be flourish again?

Let's take a look at the great changes in the world economy. Corresponding to the new economic theory and new industrial development mode of Internet economy, the economic and industrial values have changed dramatically. For instance, "deindustrialization" and "asset-light", the core values of economy, industries and enterprises are constantly flowing to the Internet industry, industrial Internet industry and financial investment industry, which is further away from the traditional manufacturing industry and enterprises. The traditional manufacturing industries and enterprises are increasingly becoming the processing industries and the vassals of Internet industry, industrial Internet industry and financial investment industry. As a result, the profits and share values of manufacturing industries and manufacturing businesses will continue to decline. Therefore, the economy and industries in the three northeastern provinces have to make their structural adjustments as soon as possible to achieve a "reborn". Meanwhile, their manufacturing enterprises should gradually transform and upgrade themselves towards the "asset-light" industrial Internet companies and new modes. Under such extremely heavy pressures, the governments at all levels of the three northeastern provinces as well as those large manufacturing enterprises must deeply understand the essence of the national policies and measures on new industrial development mode. Consequently, they should become self-reliant, rise in great vigor and achieve rebirth through reform and transformation.

In fact, these structural transformation and upgrading challenges are presented to all SOEs throughout the nation. Nowadays, no business can stay away from the tides and challenges of global Internet economy.

Table A.13 of Appendix A shows the structure of China's Top 50 enterprises in 2014. It can be observed that China's top 50 enterprises are mainly dominated by the state-owned, traditional "asset-heavy" manufacturing enterprises, traditional banks, insurance companies and real estate businesses. This is fundamental structure and reality of the current economy and industry in China. To follow the pathway of Internet economy, such structure and reality tell us that China's economy and industries still have a long way to go. We should not stay where we are and enjoy what we have achieved.

An in-depth analysis shows that a large number of "large and all-inclusive" state-owned enterprises haven't fully completed the basic process of industrialization yet, not to mention the informationization which is the another critical business development process. There is lack of core technologies and competitive high-end manufacturing products. Very few businesses have adopted the new "asset-light" industrial Internet development mode. A mode or pattern is not formed where the major industrial enterprises take the lead, followed by a large number of specialized SMEs. They meant to work closely together through the clear division of labor, mutual coordination, mutual support and risk sharing. The "Meridian vessels" are not fully unblocked. In other words, the pathway is not broad and smooth. As a consequence, this narrow mode has impeded China's rapid transition and advancement towards the market economy.

Over four decades of reform and opening-up, in spite of the low starting point, China's economy, industries and businesses have always been gaining growth. This is arguably due to the massive populations, enormous markets and massive demands. This may lead to a misperception that the economic situation is constantly improving. As a result, the contradiction of irrational national economy and industrial structure may be concealed, particularly the risks derived from the "asset-heavy" investments in all industries and trades. Thus, another key aspect of economic development has been overlooked. In the event of economic downturn, efforts should be made by the industry, business and the society to study and master the core methodology with which we can control the heavy-investment risks and deal with the overcapacity of manufacturing industries.

In fact, the core values of a modern enterprise lie in high profit, high share value, high market value, high shareholder benefits, high operator and labor income, high customer satisfaction, outstanding core competitiveness and sustainable development capacity; rather than in its large scale, high output value and high sales revenue. In particular, enterprise's core values lies in "strong and fine" rather than in "large and all-inclusive". The "strong" lies in the control of industrial standards, industrial technology system, industrial ecosystem, global advantageous resources allocation and management decision-making through data and big data analysis. The "fine" lies in the control of industrial core technologies, core-values, high-end manufacturing, and sharing investment risks. Therefore, China should strive to develop the "asset-light" industrial Internet companies. The support of Internet, 5G, software, big data, AI and other information technologies should be used to facilitate in-depth integration of the manufacturing, service, financial investment, software and Internet industries. Meanwhile, China's leading enterprises should be encouraged to adopt the "asset-light" industrial development mode. As a result, they can work with those "asset-heavy" specialized small and medium supporting enterprises to establish a new industrial ecosystem. There are a large number of benefits, e.g. promoting the adjustment of industrial structure, the diversification of risks, and the realization of investment protection. In fact, this is the methodology for settling pressures derived from economic slowdown and downturn as well as dealing with the overcapacity of manufacturing industries.

In summary, compared with the western developed countries, the structural problems of China's economy lie in the long-term impact of planned economy. First, the overall layout of China's economy is mainly dominated by the state-owned enterprises. Consequently, both the bonus and the risks generated in national economic development can't be distributed to and shared with the entire society and world. Second, China's industrial structure is mainly dominated by the traditional manufacturing. In addition, the state and society are heavily reliant on the development of manufacturing industry, which created a strong "traditional manufacturing" culture. Meanwhile, the economic and industrial structures have not been fully extended and stretched in the direction of service industry, software industry, Internet industry, financial investment industry and so forth. Third, the monolithic traditional manufacturing enterprises are still adopting the "asset-heavy" and inefficient development mode. The internal structure of a considerable number of state-owned manufacturing enterprises is still characterized by "large and all-embracing". In essence, the abilities of Chinese manufacturing enterprises to withstand the economic slowdown and downturn are extremely weak, so as their capabilities to prevent business risks.

Despite a huge amount of difficulties and obstacles, China's economy, industries, enterprises and all levels of governments must not stop reforms. Indeed, we have to find the way leading to a prosperity on our own. The sense of mission and responsibility is the driving force to identify a new method, new way and new economic theory related to China's economic and industrial development.

3 The Great Ambition of China's Economic Reform Is Not Stoppable

China's economic structure is still in a process of continuous adjustment and reform. Similarly, the construction of the socialist market economy with Chinese characteristics is still in a stage of continuous development. Such a reform process is far from completion. In terms of the global economic development pattern and trend, China's economy and industry are dealing with a new round of strong impact and challenges derived from the Internet, big data, 5G and AI economy. Indeed, it is a critical timing as the current economic reform is presented with a series of challenges. On the one hand, it is necessary to assess and judge our own national conditions, economic foundation and enterprises' status quo. A detailed analysis can be found in the previous section. On the other hand, we need to calm down and ponder over the current situation. The Internet economy will inevitably present significant challenges to China's economy and industry. If we can handle it appropriately and have the courage to move forward, these challenges may probably become a rare historical opportunity for China's economy, industries and enterprises to achieve a great-leap-forward development. A number of Internet and industrial Internet companies (e.g. Alibaba, Tencent, Huawei, Baidu, MI and SF Express) have achieved such a remarkable success in the capital and stock market in a short period of time, which show the hope of "great-leap-forward". Therefore, it is imperative to re-examine and have a deep understanding of the Internet, big data, 5G and AI technologies; their true meanings; their strategic significance to the economy, industry and enterprise development at all levels; and their impact and change on the production relations, production mode, industrial development model, economic basis and superstructure. Indeed, this presents a significant challenge to the national decision-makers as well as the pioneers and operators. This will profoundly affect China's economic, industrial, enterprise development in the coming three decades in terms of trend, planning and the big picture.

Firstly, what "Rubik's Cube" or "fairy" are the Internet, big data, 5G and AI? What changes have been made by the information technology represented by the software, Internet, big data, 5G and AI? How do they change the production relations, production mode, industrial development model and our social life? Can China realize a "great-leap-forward" economic development by promoting such a new Internet economy?

For the manufacturing industry, traditionally the enterprise is product-oriented. In recent years, through the continuous application of various types of industrial software (i.e. the informationization), the manufacturing enterprises have substantially improved their efficiency and reduced their cost. A deep integration has been formed between the manufacturing industry and the software industry. In the era of enterprise informationization, the software, as a main representative of the information technology, plays the role of production tools and productive forces. Similarly, in the enterprise Internet era, enterprises have become increasingly service-oriented in order to quickly respond to and flexibly realize the market for personalized service, product life-cycle service and customized mass production. This is actually a more advanced service and more difficult target for the manufacturing industry. Therefore, the four major links of the manufacturing industry (i.e. designing, manufacturing, management and service) need to be realized by means of the deep integration between software, Internet, large data, 5G, AI and other information technologies. As a result, the optimization, improvement and innovation of enterprise's micro-structures and external relations can be achieved. In this process, enterprises need to invest huge amount of money in two aspects, i.e. "hardware" facilities and "software" development. This will inevitably result in a heavy investment and operation burden to the enterprise. In the traditional industrial development model, the manufacturing industry makes "heavy assets" by themselves for "everything", which results in significant investment and operational risks. As a result, enterprises will be forced to actively explore and develop a new inter-enterprise cooperation method based on the Internet, 5G, software, big data and AI so as to share these heavy investment risks with others. Consequently, a new industrial ecosystem will be formed through the innovations of production relations, production mode and industrial development model.

This is essential to share investment and operation risks so that win-win mode can be achieved. In the process of exploring such a new economic methodology and achieving such a transformation, the production mode of "zero manufacturing" or "indirect manufacturing" has been implemented innovatively by the Apple, led by Steven Jobs. This has been a successful practice of new "asset-light" industrial development mode with the aim of operating the core values of industrial ecological system. A systematic explanation or interpretation has been provided for the great changes and reforms of production mode, production relations and industrial development model. This provides very useful information to the global economic sector, industrial sector, financial investment circle, entrepreneurs and macro decision-makers. Apparently, it is the starting point and the original intention of this industrial revolution based on the Internet, big data and AI. In this regard, a detailed analysis was conducted in Chapter One.

The implementation of the new production method and new industrial development model by Apple in fact presents a profound truth that the Internet is more than a technical concept. Standards, software, Internet, 5G, big data and AI have been applied and integrated in the manufacturing industry. This not only enhances the production efficiency and productivity, but also leads to a revolution in production relations, production mode and industrial development model. Therefore, "Internet" has become a critical economic concept as well as a new methodology for the economic development. Therefore, with the development of this new economic methodology and its popularization, the global economy will enter a more efficient era of the Internet economy, i.e. a more advanced stage of market economy. In other words, the Internet economy is a high-level form of market economy.

It is crucial for us to recognize these changes and take corresponding actions such as actively promoting the application and integration of the software, Internet, 5G, big data and AI; concentrating on the innovations of production relations, production mode and industrial development model. As a result, China's industries and enterprises will achieve a "great-leap-forward" development. In other words, China's economy may move directly into this more advanced development stage of market economy, achieving our dream of strong enterprises, wealthy people and powerful nation.

Secondly, what is the format shown by the Internet economy? Is the Internet economy a market economy, or a planned economy, or a economic complex format which combines the planned economy and the market economy? What is the structure of future's economy? Which parties will compete in the industries? What is the format of future's financial investment industry? Why does the Internet economy have a revolutionary impact on the economic base as well as the superstructure?

The industrial format as well as the basic concepts, rules and management systems of traditional economy which we are quite familiar with is probably like this: manufacturing industry is manufacturing industry; service industry is service industry; software industry is software industry, financial industry is financial industry, insurance industry is insurance industry, telecommunications industry is telecommunications industry, and so on. The division or the edge among these industries is very clear, and each industry is relatively independent. For example, the automobile products of Ford, GM and Toyota and the household electrical appliances products of Sony and Haier are typical manufacturing format. Wal-Mart, Carrefour and DHL logistics service are typical service format. Microsoft, UG and SAP are typical software format. AT&T, China Mobile, China Telecom are typical telecommunications format. Bank of China, Industrial and Commercial Bank of China, China Construction Bank are typical financial format. China Insurance (Holdings) Limited, China Pacific Insurance (Group) Co., Ltd. are typical insurance format. It can be observed that the business of each enterprise is in a relatively independent format. Generally the enterprise is not allowed to operate other forms of business in China. According to current national policies on economic and industrial management in China, the central government will not approve the compound development mode of telecommunications, finance, insurance and other multi-format business in one enterprise. The central government and local governments in China at all levels also willing to accept such a clear division of labor for the convince of administration such as industry management, official statistics and specific services.

As aforementioned, the service-oriented Apple Company has innovatively implemented a new "asset-light" industrial development mode in order to achieve industrial ecological core values. So, what is the format of such a new ecology correspondingly? Let's have a case study of smart-phone which is most popular in the world from the Apple Company. Results show that the Apple Company is indeed a manufacturing format. Its smart-phone products are designed by Apple Company and produced and manufactured by thousands of enterprises in its industrial ecosystem. Meanwhile, Apple Company is a service format as it sells the smart-phone products by itself. Similarly, Apple Company is high-end chip business format because its built-in core chip is designed by Apple Company and produced by other core enterprises. Apple Company is also software format due to the fact that Apple has developed, used and installed its basic software (i.e. iOS operating system), a large number of Apps and also many industrial software in the process of producing smart-phone products. Apple Company is undoubtedly an Internet business format as the company provides typical Internet services to global users. More importantly, Apple Company also belongs to the financial format which allows it to occupy the top-end of industrial ecology value chain. It warrants attention that Apple Company has in fact controlled all data related to the designing, manufacture, operation, management and service from tens of thousands of supporting enterprises and massive users within its industrial ecosystem. This is a typical "big data" industrial format. Dislike other traditional financial investment companies, Apple Company can even carry out a direct or indirect investment to those dominant and excellent enterprises within its industrial ecosystem. As Apple Company controls all the data of enterprises in its industrial ecosystem and knows the real operational situation of those enterprises, its investment efficiency and success rate will be higher and investment risk will be lower. There is no doubt that such a new financial investment format represented by Apple is bound to have a strong vitality and market competitiveness. Apparently, the financial format based on big data and AI represents the future's direction of financial investment industry. Now, you might imagine the Apple's insurance format, and so forth.

In summary, the development strategy, business, production relations, production mode and industrial development model of Apple Company actually have covered a variety of formats such as: manufacturing, services, core chip, basic software, application software, industrial software, Internet, industrial Internet, big data, AI, financial investment and insurance. This is indeed a massive industrial ecology and industrial ecosystem. More important, this is a highly complex industrial format. Meanwhile, Apple Company can secure a large amount of industrial ecology's core values. Such a new industrial format is entirely different from the traditional single industrial format which we are familiar with.

"Two Mr. Ma" is a recent phenomenon in China's capital and stock market, i.e. the large number of investments have been made by Ma Yun of Alibaba and Ma Huateng of Tencent in their industrial ecosystems. Both of them have higher investment efficiency, higher investment success rate and lower investment risk. It is not hard to imagine that, once they control the big data of their respective industrial ecosystems, they will show extraordinary financial investment attributes and strong financial competitiveness. In fact, a highly efficient Internet financial service format has emerged in China, which is represented by Alipay and WeChat-payment, and "Two Mr. Ma". This tells us that it is far more complicated than an e-business platform, a live chat platform or an App.

In addition, there is another eye-catching company in Chinese market, i.e. MI. Currently, it is the most successful industrial Internet Company in China. MI looks like a replica of Apple Company. Such an industrial Internet company will probably outperform Internet service companies such as Alibaba and Tencent in the future in terms of technology, management, services, core competitiveness, innovation and influence in the financial capital and stock market. The positioning of MI is to provide "Internet + electronic" products and services. With the digital home appliances represented by smart-phone as its entry points, MI is gradually entering into the industries of digital appliances and white goods. Its service target and business object cover all hardware and software related to digital appliances and white goods. What MI pursues is that all home appliances are interconnected through the Internet. Smart-phones, PAD, computers, televisions and other controlling platforms are set as the core. As a consequence, 24 h network management, control and service for all home appliance hardware products can be realized through various types of application software. This means that televisions, refrigerators, microwave ovens, rice cookers, air conditioning, air purifiers, washing machines, water heaters, home robots, medical and health equipment, lights, switches, locks, even cars and other electronic products are all connected to form the network and the environment of family service.

At the same time, the active value-added services of each household electrical appliance can be realized through the monitoring and accurate analysis of "big data" related to the actual operation of each individual appliance. For example, each household's energy-saving and diet habits can be interrelated to personal health; the temperature and humidity at home and automobile can be controlled in advance; foods in refrigerators can be linked to the direct logistics delivery. Similarly, the value-added services such as active pushing the information following interest by the householder and active maintenance services of home appliance can be provided. Whenever there is rigid demand and personalized need for family services, MI is able to allow the householders themselves to meet these service needs through software, Internet, 4G, 5G, big data and AI. This is fundamentally different from the business model of traditional white home appliances and information appliances enterprises for which the profit and value are realized by simply selling their products. For MI, service is the core and driving force of everything. In fact, despite seeming to provide home appliance products, MI essentially provides services, Internet active services, big data and AI consulting services and value-added services.

From a deeper perspective, MI operates a large industrial ecosystem. Such an industrial ecosystem includes manufacturing industry, service industry, chip industry, software industry, Internet industry, big data & AI industry, the financial investment industry and so on, which is also a complex multi-industry format like Apple company. A further analysis revealed that MI adopts the industrial Internet development model of "asset-light". In terms of the manufacturing operation of this compound format, MI does not participate in any sector of processing, manufacturing, assembling and testing, etc. All those sectors are outsourced to the dominant enterprises in its industrial ecosystem to share business, dividends, and investment risks, thus realizing a win-win development.

In such production relations, production mode and industrial development model, when MI sells smart-phones, PADs, computers, televisions, refrigerators and other home appliances in the market, it will inevitably have a competitive advantage with those traditional white goods and information household appliance companies, such as Haier, Midea, Gree, Changhong, Huawei and ZTE. We can boldly foresee that MI resort to its "ace", i.e. providing some home appliance hardware products to customers with much lower price or even "free of charge" in the future's market. This is because MI can obtain profits and competitive advantages through the achievement of core values in its multi-industrial format (e.g. active value-added services of home appliance functions, big data & AI consulting service and financial investment services), as well as high stock value and high market value. Simply put, what Haier, Midea, Gree, Changhong, Huawei, ZTE and other traditional white goods and information household appliance manufacturing enterprises prepare for the market is like one or two dishes for a banquet. These enterprises will never sell the products at the price lower than the cost or sell the products "free of charge". On the contrary, an industrial Internet company like MI prepares all courses for a banquet, thus can give one or two dishes to the customer free of charge. This is essentially an asymmetric business competition or industrial war. The market competition and its outcome are derived from various industrial and economic values. Meanwhile, we should be aware that an industrial revolution has been quietly taken place. Is there anyone turning down smart-phones, computers, televisions, refrigerators and other home appliances "free of charge" from MI? As long as you are willing to purchase and join in the service environment of "Internet + electronic" provided by MI, you can have some or all these products free of charge! Who will then be willing to spend money to buy other company's home appliances products?

Core values and endless values are provided by means of the integration and innovation of software, the Internet, big data, AI, with the new business, new production relations, new production mode and new industrial development model. MI and other industrial Internet companies will actually overturn those large number of traditional white goods and information household appliance enterprises which will ultimately become the specialized processing and manufacturing enterprises within the ecosystem of industrial Internet companies.

Meanwhile, in terms of the financial investment orientation of this compound format established by MI, the company has in fact controlled the "big data" of designing, manufacture, operation, management and service of thousands of matching enterprises in its industrial ecosystem. When MI makes direct or indirect investment to those dominant and high-quality enterprises in its industrial ecosystem, all the data of invested enterprises are transparent. As a result, MI will have higher efficiency and success rate of investment as well as lower investment risks. This new financial investment industry format represented by MI is bound to have a strong market competitiveness and sustainable development capacity, representing the development model of China's financial investment industry in the future. I am wondering, after gaining the control of the industrial ecology of white goods and information household appliance, maybe MI will use the same methodology and mode to enter into automobile industry, home furniture industry, real estate and other industries closely relevant to family life in the near future. This will lead to the revolution of those traditional business and industries. As a result, a bigger industrial ecology and industrial ecosystem will be formed. A large amount of core values will be obtained and achieved in the future!

It seems that in the future, the companies with complex multi-industry format will definitely replace the traditional enterprises with only single format. For an enterprise or for an industry, technological innovation is far from enough. More efforts are required to carry out multi-dimensional innovations in terms of production relations, production mode and industrial development model. Companies can survive and sustain development in the fierce market competition by keeping pace with the times as well as closely embracing software, the Internet, 5G, big data and AI. Otherwise, they will pay the price or even be replaced. This will be the pulse, voice and fate in the era of Internet and big data economy.

The above analysis shows that the Internet and industrial Internet companies (e.g. Apple, Google, Alibaba, Tencent and MI) are all showing a new compound multi-industry format and a huge new industrial ecosystem. The irresistible new mode, new model and new economy will transform the traditional service format, manufacturing format, software format, telecom format and so on. Similarly, the traditional financial industry, capital industry and the insurance industry will not be dominant anymore. Finally, the financial investment industry format and insurance industry format will be transformed. The industrial format is changing, and a new industrial revolution has begun. As a new economy format, the Internet economy is quickly approaching to us. So, what kind of economy is the Internet economy? Is it market economy, planned economy, or other forms of economy?

A huge cross-industry ecosystem is built by the Internet or industrial Internet companies. From the micro perspective, what leading companies pursue should be a "strong" planning on scheduling, delivery, logistics, index, performance, quality control, cost, standard, norm, management and service for processing, manufacturing, assembling and service of their supporting enterprises. The whole industrial ecology must be lean management and service based on the unified industrial standard, software, Internet, big data, AI platform, i.e. the "cloud". In such a huge ecology, problems or deviations of any link will affect the entire system. Therefore, the management and service among enterprises and among various industrial formats should feature with highly collaboration, transparency, standardization, networking, datamation and AI. Therefore, what such a huge industrial ecosystem pursues is bound to be a high level of "planned economy" or a "strongly planned economy" dominated by its leading companies. This is the first conclusion. For enterprises in industrial ecosystem, especially the dominant and leading Internet or industrial Internet companies, they need huge amounts of capital investment. Therefore, these dominant and leading enterprises must be highly capitalized, joint-stockized, and must be listed. Thus these enterprises must be in pursuit of a high market economy. This is the second conclusion. We can also foresee that the future's SMEs must feature with the high specialization, socialization, software implementation, informatization, datamation, Internet implementation and servitization. They could provide their services for one industrial ecosystem or for different industrial ecosystems so as to maximize their own values. From a macro perspective, future's economic society will definitely see intense competitions among different industrial ecosystems. It is unlikely that a lot of companies are able to establish such strong and huge industrial ecosystems, which reflects a high degree of market economy. This is the third conclusion. Obviously, future's national economy consequentially is controlled by such industrial ecosystems or by a small number of leading Internet or industrial Internet companies. The economy and industries will be highly centralized, which will be the new economic structure and format under the Internet economy. Surely, this new economic structure and format will be bound to change China's economic base and superstructure. At that moment, China's national economic management will mainly concentrate on managing those strong industrial ecosystems and their macro "big data" through the network. Thus, China's national economic management and service will feature with more simplification, industrial integration, high flattening, high efficiency, standardization, Internet implementation, big data & AI, scientific and precision. Therefore, China's national economic management will be becoming more feasible under an ideal "highly planned economy" based on those macro "big data". In addition, such management mode of the national economy based on macro big data is in fact indirectly realized through those leading Internet, industrial Internet companies and their industrial ecosystems. Meanwhile, it is better choice that those major targets and tasks determined by the national plans are implemented and achieved by the leading Internet and industrial Internet companies with strong market competitiveness and innovative capability. This is the fourth conclusion.

The nation will be strong when its enterprises are strong, and vice versa. In conclusion, what the Internet, industrial Internet companies and their new industrial ecosystems pursue is actually an economic entity which dialectically unifies with the high degree of planned economy and market economy. This is the essence and nature of the format of Internet economy. This might become the new economy philosophy to be discussed in the future. I am keen to carry out deeper academic research and discussion with readers.

It seemed that the human society has approached to a large industrial economy in the past based on collective capability with reliance on energy and machinery tools from a self-sufficient agriculture economy. Today, the human society is marching to an information service economy based on global interconnection and big data, i.e. the era of "brain" economy and smart economy. The development context of economy and human society has become increasingly clearer.

Thirdly, the connotation and significance of Internet and industrial Internet have been discussed in previous section. Let's revisit the traditional manufacturing industry now. What is the core value of manufacturing industry? How the core technology and core value of manufacturing are realized? Why the industrial Internet will become a rare opportunity for China's manufacturing industry and economy to achieve a "great-leap-forward" development? Is the so-called "virtual economy" a pseudo-concept?

For the manufacturing industry, we might simply divide it into the two major sections: processing, manufacturing, assembling of lower value-added general parts, components and machine shells; and R&D of high-end base and core components. We can let the vast number of specialized SMEs to complete the large-scale manufacturing tasks of those general parts, components and machine shells. Despite low technological content, low profit and large workload, this sort of work creates a large number of job opportunities. With a huge population of nearly 1.4 billion in China, this huge market plays a critical role in manufacturing sector and the social stability, which also provides strong support to the high-end manufacturing industry. Therefore, it is necessary to keep this market alive.

So, what are sources of the profits and core values of the manufacturing industry? Undoubtedly it is in the high-end base and core components, as well as their materials. Although this segment of the manufacturing industry seems small, it creates a staggering profits and is gradually becoming the core for manufacturing industry to realize its values. It has always been the commanding height which enterprises, industries and economies of various countries contend for. For example, the high-end base and core components imported from western monopoly enterprises contribute to 80% of profit of the products from China's Sany Heavy Industry and Zoomlion company. Let's take an airplane as an example. The largest proportion of profit for the manufacturing sector is the high-performance engine. In addition, the most challenging technology and the highest value for engines are the hot parts, hot components, single crystal blades as well as their processing, equipment and technologies. These have always been the bottleneck of China's aviation industry. There are many other similar cases in China's manufacturing industry. Chinese enterprises have attempted to enter the field of high-end base and core components for several times, with unsatisfactory results. As a warning for the future, we should be aware that developed countries (e.g. UK, Germany, the United States and Japan) and their leading enterprises have invested billions of dollars, a lot of human resources and experienced fierce competition in the process of industrialization during the last century. Finally, the relatively stable monopoly has been formed in the field of base and core components in each industry. For Chinese enterprises, it is apparently difficult to squeeze into such profitable markets. In fact, it is unrealistic for those leading Chinese enterprises to invest millions or billions dollars in R&D in a short term to crack such a high threshold of high-end manufacturing industry for base and core components, due to their financial capability indicators such as profit, stock value, market value, market potential and financing capacity. It seems that we must have another brick in the wall to succeed.

In recent years, the Internet and industrial Internet companies represented by Alibaba, Tencent, Baidu and MI are gaining continuous innovation and rapid development. This paves the way for the Chinese manufacturing industry to realize a "great-leap-forward" development in high-end base and core components. On the one hand, through the implementation of "big data" strategy, these companies are marching toward a deeper level of artificial intelligence in the field of information technology. With continuous breakthroughs, the experience of Internet services is constantly improved. On the other hand, through the implementation of the "Internet +" industrial development model and strategy, these companies are continuously entering the territory of real economy and traditional manufacturing industries. They are starting to explore the automobile industry, robotics industry, medical equipment and health industry, core chip industry, white goods and information appliances industry, transportation and other industries. They are making progress all the time while opportunities are presented for the transformation of real economy and traditional manufacturing industry. Investing one billion RMB and two years in R&D, MI announced at the end of February in 2017 that the "Pinecone" chip has been successfully developed for smart-phones. With the technology development platform formed by "Pinecone" core chip, it is expected that in the near future MI will develop a growing number of core chips for various home appliances so as to reach the top of the value chain of appliance industry. This will give MI a "wings of angel" for its household "Internet + electronic" products and services. Similarly, it will promote sustainable development of MI's industrial ecosystem. As aforementioned, enterprises have to invest massive funds in R&D to control the high-end base and core components of the manufacturing industry and fulfill the high threshold of such industrial technologies. This has presented significant challenges for Chinese enterprises. However, it can be observed that once the Chinese Internet and industrial Internet companies (e.g. Alibaba, Tencent, Baidu and MI) were listed, they would ultimately obtain enormous funds from the capital and stock market, thus will have great investment capacities. It is assumed that the high-end base and core components of the manufacturing industry have high profits and returns on investment. Therefore, we can expect those Chinese Internet and industrial Internet companies will seize the opportunities and make heavy investments in high-end base and core components with their "Internet +" industrial development model. They will define a new pathway for the Chinese high-end manufacturing industry.

Now, let's revisit the goals and strategies of Germany's "Industry 4.0" discussed in Chapter Three. The industrial development model of "+Internet" is to deeply integrate standards, software, Internet, big data and AI with the development strategy, business, process, management and service of the enterprise. As a consequence, the leading enterprises of traditional manufacturing industry are able to realize the standardization, software implementation, informationization, Internet implementation, big data, AI and servitization. Ultimately, these leading enterprises can transform to "asset-light" industrial Internet companies to operate core values of the industrial ecology. Once been transformed into the Internet company, those enterprises will be highly sought-after by the capital market and stock market, attracting huge amount of funds. Meanwhile, those enterprises will have confidence and financial resources to enter the high-end base and core component market. As a result, they are more likely to succeed when competing or engaging M&A with western monopoly enterprises.

In fact, a rare opportunity has been presented to the Chinese leading manufacturing enterprises. China has a large market and domestic demand. Similarly, China has become the largest core links of global manufacturing supply chain, which indicates a massive demand from the overseas market demand. As a result, a large number of Chinese leading manufacturing enterprises have already become the largest "off-line" processing, manufacturing and mechanical assembly enterprises in the world. Their existing supply chain, service chain and ecological chain are also the largest in the world. It will be crucial for these large number of Chinese leading enterprises to implement new industrial Internet development model (i.e. the "+Internet" model) through in-depth understanding of core concepts of industry 4.0. Similarly, it is critical for these enterprises to profoundly understand and apply standards, software, Internet, 5G, big data and AI; and deeply integrate these with their development strategy, business, process, management, service and industrial development model. As a result, they will achieve a high degree of integration between "off-line" and "on-line" in the end. This will provide strong support for those enterprises to realize the new production relations, production mode and industrial development model. By that time, a large number of Chinese leading enterprises will inevitably be transformed to the top industrial Internet company in their respective global industry. Thus they will be highly sought-after by the global capital market and stock market, realizing the greatest values of their own. Therefore, it also presents a rare opportunity for the Chinese manufacturing enterprises to grow from big to strong. It is the time for China's leading manufacturing enterprises, entrepreneurs and macro-decision-makers to wake up. Otherwise, they will continue to be exploited and bullied by others. Of course, the "virtual economy" advocated by some Chinese traditional manufacturing entrepreneurs is a pseudo-concept.

I believe, no matter how the "value cube" of the Internet, industrial Internet, 5G, big data and AI is changing and unpredictable, the manufacturing and real economy are always hotly contested. The reason is very simple. What the Manufacturing industry and real economy represent is the rigid demands of basic necessities for human survival and good life (e.g. clothing, food, shelter, transport, health care, education, etc.). No one can live without them. However, I need to dialectically point out that, from the perspective of political economics, the emergence and development of "Internet" will not only promote the production efficiency and productivity, but also lead to a significant shift of production relations, production mode and industrial development models. Furthermore, the current and future businesses of those leading enterprises in China and overseas will show multi-industry formats, covering various industries such as manufacturing, service, chip, software, Internet, finance and insurance. As a result, these leading enterprises may obtain the core values from multi-industry formats, rather than the local value of the manufacturing sector. Therefore, the revolution to manufacturing industry and other related industries has already started. Furthermore, Internet and industrial Internet companies simply regard the manufacturing industry and real economy as carriers to realize their values. There have been dramatic changes to the economic and industrial values. Even the philosophy of economic development will change inevitably. From the perspective of such new value of global economy and industry, manufacturing industry is merely "senior workers" or "senior employees" rather than the master and leader of economy, industry and business. In addition, such rules have been historically transferred to the Internet and industrial Internet businesses.

For "smart manufacturing", according to this new economic and industrial values of the Internet, big data and AI, it is a valid statement no matter how intelligent manufacturing equipment is, no matter how much enterprises have spent on purchasing such intelligent equipment, and how much enterprises have spent on building their magnificent unmanned workshops and factories, the issues of over-capacity, and significant operation and investment risks can never be solved in the event of oversupply derived from the over-competition of "products". Thus, what the manufacturing industry pursues should be the deep understanding of the "market" and the "service manufacturing" rather than simply the "smart manufacturing". Everything must be service-centric, and everything must be directed to the "services" of the market and customers. Nowadays, we should recognize such major principle of Internet new economy and such a new value of economy. Therefore, we must devote to developing our own service-oriented Internet and industrial Internet companies. As a result, we can occupy the top of world's new economy and achieve the Chinese dream of "strong nation, strong enterprises and wealthy people". This will also become the final strategy, plan and blueprint for China to realize the economic restructuring, transformation and upgrading of industries and enterprises, supply-side reform and new Internet economy.

Fourth, under economic tide of Internet, big data and AI, what are the future variables of China's economy? What is the established economic policy of the United States for the global economy, industries and technologies? How should we treat the economic variables of the United States, that is, the "US priority" and "return of manufacturing industry" strongly promoted by the President Trump?

The development and reform of state-owned enterprises in China have been analyzed and discussed in detail in the first two sections of this chapter. Due to the large amount of capital investment required, the development of China's Internet industry has been highly demutualized and market-oriented from very beginning. Meanwhile, it is with high degree of Internet implementation, big data & AI implementation and service-orientation. These have also been discussed in previous sections. Discussions on the development and changes to China's state-owned economy and the Internet industry will not be repeated in following sections.

Now let's briefly discuss the largest, most dynamic and promising economies in the current Chinese economy, i.e. private economy, private enterprises and township enterprises. Since the reform and opening-up in 1978, China's economic and industrial development has experienced fluctuations. Nowadays, a large number of private enterprises and township enterprises have played a critical role in China's real economy and manufacturing industry. This achievement cannot be made without the dedication of a large number of hard-working entrepreneurs of private enterprises and township enterprises. They deserve our respect for their indomitable entrepreneurial spirit. A very high proportion of these private enterprises and township enterprises are family business. Along the golden age of rapid development in last four decades, a large number of those entrepreneurs are now at their sixties and seventies. Historic challenge is presented to these enterprises, i.e. gradual "entrepreneur replacement". Basically entrepreneurs of those private enterprises and township enterprises realize their entrepreneurship and development by doing everything by themselves. Generally speaking, they do not have systematic knowledge on enterprise management or profound understanding of modern enterprises. As a result, they inevitably believe in their own experience and practice. They tend to be conservative for new things such as new information technologies, new production relations, new industrial development modes and new economy. As a result, their enterprises may lose the golden opportunity for the transformation. On the one hand, a large proportion of family enterprises with oriental culture (e.g. Japan, South Korea and Southeast Asia) struggled due to failing to deal with the challenge of "entrepreneur replacement" properly. Obviously, China's private enterprises and township enterprises will sooner or later deal with the similar challenges. The proper solution lies in the full understanding and reference to the corporate governance structure, professional managers and other modern enterprise systems adopted by enterprises in the United States and other western countries. On the other hand, China's entrepreneurs of private enterprises and township enterprises are very familiar with the business of real economy and manufacturing industry. However, they have very little knowledge on the software, Internet, 5G, big data, AI and other applications of information technology. These entrepreneurs have to deal with the second entrepreneurship challenge when facing the increasingly overwhelming impact of the Internet, 5G, big data and AI. We can see that those entrepreneurs are no longer young. It is far exceeded their ability and power. As a result, these enterprises may miss the opportunity for the transformation and upgrading, and probably fail in the fierce market competition later.

In fact, when dealing with the challenges of Internet and big data economy, China's private enterprises and township enterprises should actively adjust their traditional production relations, production modes and industrial development models. This can be achieved through the supports and deep integration of software, Internet, big data, AI and other information technologies. As a consequence, a "combined fleet" can be established with related enterprises where those leading private enterprises and township enterprises play the role as "aircraft carrier". Only this way can they develop the capacity and market competitiveness. The reason is very simple. This is the age of data concentration for enterprises, industries and industrial ecology, i.e. the age of big data.

I believes that the market and the economy are in a state of changing. It will be a variable and a threshold of the future Chinese economy how Chinese private enterprises and township enterprises are able to cope with the double challenges of the "entrepreneur replacement" and the era of Internet economy. This warrants a high degree of attention from the society and all levels of governments. Private enterprises and township enterprises must be vigilant and advance with the times. Only this way can those private enterprises and township enterprises realize their sustainable development, and support the half of China's economy and industries.

Next, let's discuss the variables of the U.S. economy. I believe that there are actually two dimensions of economic policies in United States. The first dimension is the economic policy developed and implemented by the U.S. federal government. The second dimension is the economic policy that is consensus reached by the financial magnates of Wall Street in New York, the industrial circles and the Silicon Valley's high-tech communities. This is often confused and must be clarified. According to its constitutional structure, the United States indeed has a small administrative government and a large society. The United States implements a highly competitive market economic system. Its core is Wall Street's high control over the finance to the industries and their core technology of U.S. and over the world. In addition, the U.S. federal government mainly relies on the fiscal, tax and other policies to regulate the market and the economy. Its fiscal policy has limited impact whereas its tax policy is relatively more influential. In general, the U.S. federal government's economic policy has limited capacity to control and regulate the market. The implementation of US economic policy is mainly accomplished and realized through the financial giants of Wall Street. Therefore, the United States will have a huge impact on the global economy if the federal government can reach a consensus with those financial giants of Wall Street in economic and industrial policy. Otherwise, the impact of United States on the global economy and industry will be limited even though the federal government unilaterally makes strong implementation of an economic or industrial policy.

During last three decades, the United States has been leading the world in the development of information technologies such as computers, communications, core chips, software, Internet, big data and artificial intelligence. Nowadays, the United States is also leading the world in terms of the Internet economy, big data economy and new industrial development mode. The Wall Street through the control of financial industry has long been made a consensus with main industrial circles and the Silicon Valley's high-tech communities to support Internet industry, soft industry, high-end manufacturing industry and service industry to realize high profit and high market value. These industries are encouraged to vigorously develop "Asset-light" Internet and industrial Internet companies. These have helped the U.S. and its companies to always occupy the top end of the global industrial value chain.

It is worth noting that the U.S. federal government has reached a consensus with the financial giants of Wall Street, the high-tech elites of Silicon Valley and various industries while ago on the future planning and layout of the global economy, industry and technology. According to this plan, US is the "brain, nervous system and heart"; Germany, Japan and other western developed countries are the "heart, liver, spleen, lung and stomach"; and China, India and other developing countries are the "trunk and limbs". This is the essence of the global industrial revolution advocated by the United States. In other words, the United States occupies the top of global industrial and technological value chain. As a result, the United States will seize the control of global industries, real economy and market through the Wall Street's finance, industrial standards, intellectual properties, as well as Silicon Valley's software, Internet, big data and artificial intelligence technologies. Germany, Japan and other western developed countries will occupy the middle-end of global industries and economy. On the contrary, China, India and other developing countries will occupy low-end industries and services. This, of course, has raised concerns from Germany, Japan and other western countries. Germany is attempting to deal with it through the implementation of "Industrial 4.0" strategy. Thus, when the former President Barack Obama proposed the "re-manufacturing" economic policy that aimed to solve the unemployment issues, there was actually no active response from the financial giants of Wall Street, high-tech elites of Silicon Valley and various industries. As a consequence, that policy basically had no impact on the global economy and industries.

So, those three powerful U.S. parties (i.e. the financial magnates of Wall Street in New York, the main industrial circles and the Silicon Valley's high-tech communities) may not necessary agree with the economic policies of "U.S. Priority" and "Return of Manufacturing Industry" rolled by the President Trump. They may not provide any actual support. If the President Trump tries to vigorously promote this new economic and industrial policy, it will definitely lead to conflicts between the U.S. federal government and the financial giants of Wall Street, high-tech elites of Silicon Valley and various industries. The term of U.S. President is only for four years or eight years at most. Therefore, conflicts or inconsistency between two dimensions of main economy policies will inevitably result in confusions on the economic values of the United States, which in turn lead to the chaos of the economy and industry of U.S. and the world. As a result, the finance, economy, industry and their enterprise of the United States will suffer from low efficiency, high cost and low global market competitiveness. Obviously, this is a variable for the future's U.S. economy. If China can be stabilized at this critical period, it will surely be a rare opportunity to catch up with the U.S.

Surely, today, whether it is the United States, China, or even the world, the largest economic variable is the Internet, industrial Internet, 5G, big data and AI. For nations, economies, industries, enterprises and even individuals, the Internet, industrial Internet, 5G, big data and AI are the most core values in the future.

In summary, we have revisited the true meanings of Internet, big data and AI technology as well as their derived new economic philosophy and value. To appropriately deal with the current critical issues associated with the economic restructuring, supply-side reform and industrial transformation and upgrading, and for achieving the blueprint of developing Internet economy in China, I believe three strategic base points should be discussed and mastered.

First strategic base point: We should fully understand China's national conditions and continue to give full play to the advantages of our system, that is, the planning advantages of the central government to concentrate on resources and strength with major events. This is actually derived from the planned economy. Various strategic industries and products of China have become competitive in the fierce market competition globally in such a short period of time. These include: high-speed railway, freeways, large bridges, electric vehicles, large aircraft, satellites, launch vehicles, high value ships, shield machines, UHV power grids, ultra-supercritical power generation, supercomputers, mobile communications and other major national economy infrastructure. This is because of the appropriate decision-making and policy guidance by the strong planning, management and centralized R&D support of the state; and the insistence and efforts of a large number of state-owned enterprises. China has only developed the market economy for around three decades, which is still in the initial stage. While, western economic powers and their leading enterprises have spent hundreds of years to develop the market economy. Therefore, they have great advantages and create enormous pressures on China. It is unrealistic and infeasible for Chinese enterprises only by themselves to catch up with western leading enterprises in the market by means of free competition and natural development. Therefore, our advantage of concentrating on resources and strength with major events must be maintained in China. This is a base point and also an old tradition, which should not be discarded at will.

China's high-speed railway system and EMU train have become more competitive in the international market, which make a large number of Chinese people convenience to travel. This has revitalized the Chinese regional economy and have also started to serve the strategic plan of the "Belt and Road". Such a achievement should be attributed to the strong programming, planning and policy support of the state, the Ministry of Railways, and relevant governmental ministries. This should also be attributed to the hard working and dedication of scientific researchers, workers and the employees of CRRC Corporation Limited. Imagine if, according to the development law of the market economy, private enterprises gradually accumulate funds, technologies, and capabilities, how long the public has to wait to enjoy the convenience brought by high-speed rail and EMU. How long it will take for the nation to make this industry to be the core industry. The fact is that at the beginning, China's high-speed railway and EMU technologies were half century behind that of Japan, France and German. In addition, the investment on high-speed railway and EMU should be as high as hundreds of billions or even trillions of dollars. Thus, the national conditions, industry status and enterprise status should be carefully taken into consideration. Such vast and arduous engineering works must be guided and implemented by the "strong" planning of the state.

In fact, the western developed countries and their industrial giants have realized strong market competition advantages through hundred years of technology accumulation and industrial development. Therefore, it is absolutely necessary to maintain the "strong" planning support of the state so that the Chinese core industries and high-end products can develop competitiveness advantages in the international market. This is a reality.

At present, some Chinese economists are worried about the planned economic system which has once been implemented in China, and even hate it. They believe such economic system is inconsistent with the development trend and values of the international economy due to the lack of market competition. As a result, the quality, innovation and competitiveness of the national economy, industry and enterprises will be affected, which also results in the structural problems of the Chinese economy. Such views and ideas are understandable. However, the value and management of economy are also the dialectical unity of "changed" and "unchanged". The "planning" element in economy is the eternal core value. The immutable economy system is actually rigid, inflexible and unsustainable, as the market may fail in some occasions.

Let's take a look at the original intention and significance for the European countries to set up the European Union (EU). All western European countries implement the market economy system and encourage the free competition. At the beginning, each western European country fought in the market for itself and developed on its own. However, each of them is relative small in size and weak in power in terms of the population, market scale, economy, industries, resources, technologies and so on. There might be variables and changes in market competition, e.g. the growing competitive pressures and challenges from the United States, Japan, and their enterprises. In such circumstance, these western European countries and their enterprises would feel powerless and lose ground in the market. Reflecting from the painful experience, they began to be forced to explore new production relations and industrial development model. They established a new economic base and superstructure for which they spontaneously organized and set up the European Union, the European Parliament and other unified institutions. EU member states have to rigidly pay their fiscal budgets to the EU in accordance with the well-defined plans. The EU has adopted the unified market rules, laws, policies and all kinds of strict plans such as scientific research and innovation plans. Within the EU, the sharing of markets, finances, technologies and resources could be realized among member countries, industries, enterprises and human resources by means of "huddling together for warmth". As a consequence, the cost of each country and enterprise can be significantly reduced, and the production efficiency of an enterprise can be substantially increased. The external competitiveness is improved significantly. In addition, one of the prominent features is the emphasis on the unified planning of the EU. So, it seems that the factors of "planning" are indispensable even in such a well-developed market of European countries.

Therefore, dialectically speaking, there are always merits of planning. "Planning" and "market" should be a dialectical unity, which should be made according to the national conditions and reality. The elements of planning in the economic system cannot be discarded at will.

Second strategic base point: it is imperative to deal with the relationship among state-owned enterprises, private enterprises and foreign enterprises in China's economy properly. The dominant role of market in the national economy should be fully reflected, the power of market economy should be fully taken advantage of, and the market order should be regulated through competition.

For those enterprises involved in the core industries or national security, their majority ownership should be controlled by the leading state-owned enterprises; if necessary, some private enterprises and social capitals can be attracted to release certain part of the economic reform's dividend. Thus, the market vitality of private enterprises can be integrated. Similarly, by using the social supervision mechanism of private enterprises and the society, the leading state-own enterprises can have healthy development based on the economic rules of the market. Finally, the mixed ownership mechanism is formed, in which the state-owned holding is the main body while the private and social capitals coexist. At the same time, corporate governance structure could be introduced such as professional managers to business management. Similarly, modern enterprise system could be adopted to guarantee the constant appreciation of state-owned assets as well as interests of shareholders. This should be the best policy for the structural reform of existing leading state-owned enterprises.

Other industries and sectors fall within the scope of full market competition. The state should orderly open those markets, and the state-owned enterprises should gradually retreat from these industries and sectors. The reform dividends and investment risk of economic development should be released to and shared with the private enterprises and the entire society around the world. The purpose is to give full play to the market strength, social capital and international capital. As a result, China's economy, industry and enterprises can be fully developed by combining with the strength of the entire society around the world.

For the foreign enterprises, a good policy and healthy market environment should be developed to encourage them to enter into the Chinese market. If these foreign enterprises gain development in the Chinese market with benefits, those Chinese enterprises can identify their own problems, gaps and deficiencies in the competition, which will motivate them for more reform, development and innovations. Meanwhile, the nations behind those foreign enterprises can gradually become our friends and "allies" of win-win cooperation in the international arena. Furthermore, the grand strategy of the "Belt and Road" requires such tolerance and strategic vision. To gain development and be stronger in the future, China must be a country that is highly open and respectful of diversity. The world needs to develop together to share the glory.

Recently, there are some various opinions among "moguls" on the Chinese economy about the full marketization and moderate interference by government's planning. I would like to talk little about this issue. Let's take a look at the "1997 financial crisis" of Hong Kong. The powerful western financial predators used the management loopholes of Hong Kong market economy policies to launch the war of financial storm. The whole Hong Kong was faced with further huge losses. In such critical timing, the Hong Kong SAR Government made decisive actions with the strong support of the Chinese central government. These decisions were proved to be effective to protect the stability of Hong Kong financial and stock market. In other words, the government directly dealt with this financial crisis. If the Hong Kong SAR Government at that time emphasized the full market economy rather than the moderate or strong intervention of government "plan", the financial and stock markets as well as the economy, industries and individuals of Hong Kong would struggle and pay a big price. If a government still holds the "zombie" of market mechanism and prays for tomorrow when the market mechanism has become a complete failure, what is the reason and necessity for such a government to exist? It seems that the market and market economy are not always the universal "medicine" to solve economic and industrial development problems.

Let's take a look at the serious global financial crisis in 2008 triggered by the bankruptcy of Lehman Brothers from the Wall Street. The collapse of Lehman Brothers led to the crisis of its investors and bondholders, which caused the crisis of other financial investment companies whose financial leverage was extremely unbalanced. This further led to the crisis of the large real estate companies such as Freddie Mac and Federal National Mortgage Association, then the crisis of the majority of financial institutions on Wall Street which are the economic lifeline of the US, and the local banks. Once the crisis flooded, the industries and service industry of the US also suffered. In fact, the crisis had spread to many large American industrial companies, some of which had sought for the government's bankruptcy protection. The reason was very simple. All those leading enterprises ran out of money, as there was no cash flow. The market had become completely ineffective. At that time, the United States, as the largest and most developed free market economy in the world, no longer accused the direct intervention of Hong Kong government against the financial crisis. The US federal government and the central bank took direct intervention and action to issue bonds and print banknotes which were then loaned to other banks and financial companies of the United States. As a result, the potential financial crisis and industrial crisis which could be even worse for the United States and the world were resolved. This is a typical case of "strong" government planning intervention occurred in a so-called highly free economy. The United States had in fact issued huge amount of treasury bonds and bills in response to the 2008 financial crisis. In order to leverage these huge amount of treasury bonds and the negative effects of bills on the market, it can be predicted that the U.S. federal government and the central bank are bound to implement a medium to long term "plan" to solve this major issue and problem in an appropriate manner. This may even involve the withdrawal of state-owned shares in industrial giants such as General Motors and Ford Motor Company. In this case, what will the U.S. federal government and the central bank implement in such circumstances and in such a long period of time? Will it be market economy or planned economy? Finance has always played the critical role in the US economy. In the events of a major financial crisis and market failure, the US federal government introduced its direct intervention. Was there any nation or international organization that accused the U.S. federal government's approach of dealing with this financial crisis?

Now let's revisit the serious economic crisis that erupted in the United States in the 1930s. At that time, the US federal government also acted in its own capacity. In order to solve the significant unemployment issue, the US federal government supported a large number of American enterprises by means of large-scale infrastructure developments under the government plan. A series of policies and measures were issued to encourage social capitals to actively participate in those infrastructure developments. In the end, the US survived from that economic crisis, and an infrastructure foundation was laid for the subsequent U.S. economic and industrial development.

As a matter of fact, in the event of a major crisis, the main principle and philosophy are to survive and then to develop. This principle applies to a nation, economy, industry, business and people.

Let's revisit the impact of "open-door" policy on the automotive industry since China's reform and opening up in 1978. China's automotive industry had late start, with weak foundation and poor technology. After the Chinese automotive market was opened to the international market, those automotive production giants from Japan, Germany, France, and United States have flocked into China. What we have expected is to use the Chinese market as the price to learn foreign automotive technologies. However, we did not learn that much automotive technologies. Rather, vast majority of the market and its profits were taken by the western automotive enterprises. Even in the Sino-foreign automotive joint venture, the Chinese party had low profits and was never in the driver seat. Today, when we are walking on the street, roads are filled with cars of foreign brands. China's own-brand automotive enterprises are still struggling in the market. They struggle to survive, not to mention to further develop. The result is endless traffic jams and environmental pollution derived from exhaust emissions of automobile. This situation is unexpected and has become a mess. It is difficult for the nation and industry to deal with and difficult to find medicines to "cure". This has been criticized by many scholars and people. The market economy is a "good medicine" to develop the economy. However, it will not be appropriate if the market could not be controlled by a nation's economy and its capability. Once the so-called market has become a derailed "wild horse", such market economy will soon become a "bitter medicine" with good excuses.

No matter a market economy or a planned economy, the development and application of economics and economic theory cannot be pedantic. Indeed, it needs to make the most of situation. Obviously, the situation in which the "plan" intervenes the market will inevitably occur frequently in China's economic management at all levels of governments. It is always a significant challenge to keep the balance. Similarly, the protection and restraint by laws, regulations and systems are required.

Third strategic base point is to vigorously develop the Internet economy:

Today, we are dealing with the significant impact of the rapid development of Internet, industrial Internet, 5G, big data and AI. This is the case no matter from the perspective of global economic development or the Chinese economic development; no matter the format change from the perspective of traditional manufacturing industry, traditional service industry, or traditional financial investment industry; no matter from the perspective of innovation on technologies or on production relations, production mode and industrial development model. Indeed, the market and economy are changing all the time.

In such a fast changing era, at the macro level, we have to make judgement properly, and highly focus on the changes derived from Internet, industrial Internet, 5G, big data and AI to the economic theory and economic development methodology.

Similarly, we have to pay high attention to the significant challenges derived from the Internet economy on the Chinese traditional manufacturing industry, service industry, financial industry and other formats. As a consequence, we can cope better with the changes to enterprise form, industrial format, economic structure, economic base and superstructure derived from this new economy. We should set up a new economic value. Similarly, we should re-examine and profoundly understand the true meaning and strategic significance of Internet, 5G, big data and AI as well as their rapid development to the nation, economy, industry and enterprise. Consequently, we should formulate a practical road-map for the development of China's future economy by integrating national conditions with the foundation of industries and enterprises. Therefore, we need to design industrial development methodologies by incorporating the new theory of Internet economy whereas industrial formats need to be redefined. The production relations and production modes should be constantly adjusted and innovated based on the Internet, 5G, big data and AI. Similarly, the two new industrial development models (i.e. "+Internet" and "Internet+") should be promoted. We should encourage to change the current production relations and national administrative management systems from relatively independent single industrial format to multi-dimension integrated industrial formats. In such new industrial formats, an enterprise may cover various industries such as: manufacturing, service, core chip, software, Internet, financial investment and so on. Therefore, we should encourage to develop the Internet and industrial Internet companies aiming at operating the core values of mult-idimensional industries. In the future, we will increasingly rely on those Internet and industrial Internet companies with strong market competitiveness and innovative capability to carry out the major tasks and objects under the national plans. In the future, all levels of governments, entrepreneurs, innovation practitioners, investors and economists should maintain a high degree of coordination. All levels of governments need to constantly improve and reform the laws, regulations, systems, policies and the environments to facilitate the development of Internet economy. This can be achieved through the corresponding adjustments of the economic base and superstructure in China.

In the Internet and big data economy, it is foreseeable that the administrative management systems of national economy in the future will present the highly industrial integration, highly flattening, higher efficiency, lower cost, standardization, networking, big data and AI implementation, precision and scientification. At the same time, the Internet economy will gradually show its features or natures of both market economy and planned economy, which is a dialectical unity of planned economy and market economy. Therefore, the administration of China's economy should aim for a high degree of dialectical unity of the market and the plan. In other words, it should be clarified which or what should be controlled, be managed, be coordinated, be supervised, be reformed, be completely marketized and be fully opened. Economic restructuring, transformation and upgrading of industries and businesses, and the supply-side reform can be achieved in China through the reform, innovation and the development of Internet economy. Consequently, a "great-leap-forward" development of economy, industry and enterprise can be realized, fulfilling the "China Dream" (i.e. the rejuvenation of the Chinese nation).

In such a changing era, at the micro level, it is necessary for enterprises to deeply apply and incorporate information technologies represented by the software, Internet, 5G, big data and AI. Similarly, it is crucial for enterprises to transform from the innovation of technologies to the innovation of production relations, production mode and industrial development model, finally forming a new industrial ecology. This is essential for those enterprises to realize a fundamental change from the product-orientation to the service-orientation. As a result, enterprises can be responsive and achieve the personalized service, customized mass production and life-cycle service of products in a flexible manner.

It is worthwhile learning from practices of the Third Industrial Revolution advocated by the United States. It is necessary to formulate new economic and industrial policies so as to accelerate the opening of those Chinese industries and sectors that are still in a traditional industrial development mode (i.e. single format). Through the implementation of "Internet +" model, those Chinese Internet and software companies should be encouraged to enter into the real economy, traditional manufacturing, service, financial industries and so on in a continuous manner. These Internet and software companies should be transformed to "asset-light" industrial Internet companies with an aim to operate core values of multi-dimensional industrial formats. By adopting such new production relations and production modes, the quality and efficiency of the Chinese industry and companies can be significantly improved. Ultimately, China's global competitiveness can be enhanced in terms of economy, industry and business.

It is worthwhile learning from the practices of Industry 4.0 (German). It is imperative to encourage and vigorously promote those Chinese enterprises of traditional manufacturing industry, service industry, financial industry and so on to deeply integrate with information technologies such as software, Internet, 5G, big data and AI. Through the implementation of "+ Internet" model, the Chinese traditional manufacturing industry, service industry, financial industry and their leading enterprises can be developed into "asset-light" industrial Internet companies with an aim to operate core values of multi-dimensional industrial formats. Similarly, the current traditional production relations, production modes and industrial development model should be reformed and innovated. As a result, the quality and efficiency of those Chinese enterprises can be significantly improved, and the cost of those enterprises can be dramatically reduced. These enterprises will be transformed from the product-orientation into the service-orientation.

Therefore, China's state-owned leading enterprises should adopt a "stripped" strategy for which their large number of low & medium-end production, processing and assembling businesses and service links should be gradually separated from their parent company. Consequently, a certain number of relatively independent and specialized enterprises will be formed which can cooperate with their parent company through approaches such as mutual shareholding. These new and "stripped" enterprises should be highly specialized and can provide service to not only their parent company and but also the entire industry around the world. In other words, the state-owned leading enterprises should transfer their non-core technologies and businesses (e.g. processing, manufacturing, assembling and low-value services) to a large number of SMEs within their industrial ecosystems. By adopting such mode to divide profits, the efficiency of leading enterprises can be greatly improved and their costs can be substantially reduced. More importantly, the massive investment and associated risks can be properly shared. Finally, the new model of "win-win" and mutual development can be realized. At the same time, the state-owned enterprises should turn their focus on the research, development and innovation of core technologies as well as the high-end manufacturing industries, high-end products and industrial standards. They should pay more attention to the deep integration of Internet, 5G, software, big data, AI and other information technology. They need to place more focuses on developing the industrial technology systems with industrial supply chains and service chains. Based on the allocation of superior resources at the national and international level, new industrial ecosystems will be formed. Finally, those Chinese leading enterprises can be gradually transformed to the high-end manufacturing enterprises, core technology enterprises, industrial standard enterprises, software companies, big data and AI consulting companies, Internet service companies, financial services companies and group management companies in their respective industries or sectors. Finally, those leading enterprises will be upgraded to "asset-light" industrial Internet companies with an aim to operate core values of the industrial ecology. They will stay at the top of industrial value chain, keeping up with the new Internet economy.

In short, challenges and opportunities co-exist for China's economy, industries and enterprises following the global development trend of Internet, industrial Internet, 5G, big data and AI. The key lies in the attitude, determination and actions. This is essential to the restructure of economy and industries, the transformation and upgrading of industries and enterprises, the reform of the supply side as well as the great-leap-forward development of China's economy, industry and enterprises.

The rapid development of Internet, 5G, big data, AI, block-chain and the changes of global economic and industrial values were revisited in previous section. Based on aforementioned three strategic base points which were corresponded with the planned economy, market economy and Internet economy, If we can respectively take advantages of these three strategic base points whereas avoiding their disadvantages, i.e. creatively give full play to the advantageous "elements" among the planned economy, market economy and Internet economy in China's future economy whereas dealing better with their shortcomings and disadvantages through our economic reforms and practices, a new economic system, new economic structure and new economic development mode will be creatively developed and formed. This is entirely different from the economies of western developed countries, as they never have experiences of implementing the planned economy. Such a thought or strategy will provide the cutting-edge to China's economy, industries and enterprises, which makes the nation even stronger. This should be the China's future economic development mode and value proposition. I would like to boldly define this economic thought or strategy as the e^3 theory of China's economic development. "e" stands for the first letter of "economy", and "e^3" represents the product of the advantageous elements of planned economy, market economy and Internet economy. This is a metaphor for the prosperity of Chinese economy in the future.

For a long time and even so far, the Internet, 5G, big data and AI generally have been more regarded as information technologies or technology tools which can greatly enhance the productivity of enterprises, industries and economy, i.e. the concept of productivity force. However, my view is that the Internet, 5G, big data and AI today and in the future should be more regarded as key economic and industrial development methodologies to adjust and transform the production relations, which is more important for the coming economy, industries and enterprises. When the productivity force and its change are in accordance with the production relations and their adjustment, the enterprises, industries, economy and even the society can be quickly developed, and vice versa. So, the "e^3" economic theory is more regarded as a theory which can push to adjust and transform the production relations of enterprises, industries and economy based on the Internet, 5G, big data and AI. An advance productivity force needs and calls for an advance product relation, and an advance product relation can promote and guarantee the development of productivity. Indeed, this is a basic rule for the development of economy and industry.

A better implementation of the grand strategy of "e^3" economic theory helps to keep up with and lead the global Internet, 5G, data and AI economy. I believe that work and preparation should be made in following aspects.

First, the legal environment of economic base and superstructure should be adjusted and reformed. It is necessary to formulate the legal system, provisions and institutions that are compatible with the "e^3 economy". By in-depth application of the Internet, 5G, big data and AI, we should create a legal environment that is beneficial to the change and reform of superstructures, economic base, production relations, production mode and industrial development model.

Second, the Chinese superstructure and state administrative system should be adjusted and reformed. With the support of Internet, 5G, big data and AI, the relationship between the central government and all levels of local governments will be highly flattening and manageable. In other words, in the future, the central and state administration can significantly reduce the middle levels of local governments. Therefore, it is necessary to establish such administrative structure and system in which a central government with big data oriented "strong" planning leads a minimal level of reliable local governments. Thus the administrative costs of the whole nation can be substantially reduced and the management efficiency of the government can be significantly improved. As a result, the nation's management and services can be realized with the better transparency, datamation and more scientification. To this end, in the future, a grand "project" should be constructed for a highly unified nation-wide "standard and digital standard" of the national economy and social development. On this basis, with support of the high-speed broadband network and 5G, the central government should make massive direct investment to establish a nation-wide "State Administration and Service Big Data Center" that combines vertical and horizontal matrix. Such data center should cover all the administrative and service agencies as well as all the provinces, cities, counties, townships and villages. Meanwhile, the "Center" can be directly managed and controlled by the central government. Point-to-point and end-to-end management and services can be achieved across the nation with the high level of efficiency. The traditional practices and administrative systems were featured with networks and information systems that were built by various levels of administrative agencies and local governments themselves. Under the new mode, those traditional practices and management systems should be completely changed, which will for sure lead to substantial reduction of costs for the nation-wide administration and operation. Similarly, the efficiency of national administration and service will be significantly improved, so as the transparency, symmetry and the accuracy of "big data" of the nation.

For example, the current national statistics are carried out by reporting of local governments level by level. Such approach should be replaced by the real-time big data management based on "the State Administration and Service Big Data Center". As a consequence, the accuracy, timeliness and capacity of nation's administration and service will be dramatically enhanced for the economy, industry, enterprise, society and the emergency. Thus, the nation's macro decision-making ability, response capacity and predictability for some certain "events" will be greatly improved. Moreover, those management and services concerning people's livelihood and national services (e.g. medical care, medicine and medical insurance) should be based on the real-time big data management of the "State Administration and Service Big Data Center". Meanwhile, all the internal and external administration and services based on Internet, 5G, big data and AI should be carried out and achieved. Finally, the transparency of administration and service by big data can be realized. Consequently, the efficiency of national administration can be greatly improved whereas the costs of national administration and people's livelihood can be significantly reduced. Similarly, the life quality and happiness of citizen can be improved.

As a matter of fact, it is closely relevant to the economic security of nation and the privacy of people to determine which part of the "big data" should be protected and which part of that could be made available to the public. Its essence is the lifeline, wealth and core value for the survival and development of a nation and a people. It must be managed and maintained by the national will and citizens. Obviously, in the future, the "safety and security" of network, information and data will inevitably become the top priority that requires the attention of the nation.

Third, the nation's management system and policy on the economy and industries should be adjusted and reformed. In the future, the core of economic and industrial activities is the Internet industries and industrial Internet industries. The network service must be at low-cost. The cost of Internet and its service in fact are the cost of all enterprises, industries, national economy and the whole society.

In the future, the Internet businesses and industrial Internet businesses may cover various formats such as the manufacturing, services, software, chip, big data, AI, finance, insurance, transportation, logistics and so on. Therefore, it is necessary to change and reform the current economic and industrial administrative systems under which each of existing administrative ministries or agencies manages and controls only one industrial format. Similarly, it is imperative to establish a unified economic and industrial management system that covers all industries and services such as finance, insurance, stock, securities, telecommunications, manufacturing and services. Meanwhile, the cost of network service provided by the telecommunication industry should be strictly controlled. The innovation of production relations and production modes should be encouraged based on the standards, software, Internet, 5G, big data, AI and even block-chain.

Furthermore, the new industrial development models of both "+ Internet" and "Internet +" should be promoted to vigorously develop the "asset-light" industrial Internet companies with an aim to operate core values of the multi-industry ecology. This is essential to deal with the competitions and challenges from those Internet giants and industrial Internet giants of the United States and western developed countries. In both international market and domestic market, the future's competition will be among those strongest industrial ecosystems. This in fact represents the competitions of the enterprise capacity, industrial capacity, financial capacity and the national strength. The survival and development depend on both cooperation and competition. The "planning" or "market" orientation of the main economic policy should be determined by the national objectives, tasks and the actual needs of corporate development. In the end, a high degree of dialectical unity can be realized.

Fourth, the state-owned enterprises and central enterprises should be adjusted and reformed. Over the years, the state-owned enterprises and central enterprises have always represented the foundation and capacity of manufacturing industry and other industries in China. They reflect the core values of China's economy and industries. Today, if the state-owned enterprises and central enterprises cannot realize the reform and transformation following the Internet and big data economy, they may be retreated as the processing industry, assembly industry and low-end service industry in the future by those Internet and industrial Internet companies. If so, they cannot achieve the constant appreciation of state-owned assets, and will probably become the barrier to the national economic and industrial development. Therefore, China's leading state-owned enterprises and central enterprises need to make efforts to innovate their production relations and production mode based on the software, Internet, 5G, big data and AI. As a consequence, they can transform themselves into the "asset-light" industrial Internet companies with an aim to operate core values of the multi-industry ecology. Similarly, their competitiveness and innovative capability can be significantly improved to keep up with requirements of the Internet economy era.

Under the existing management system of state-owned enterprises and central enterprises, their managers have various levels of ranking and tenures. This means they tend to seek for short-term performance with short-term behavior. It is difficult to encourage them to make long-term development plan and implement long-term innovation activities. In fact, it will take five years, ten years or even longer for the state-owned enterprises and central enterprises to develop into industrial Internet companies. This process generally involves huge amounts of investments and relatively high level of risks. Therefore, such substantial reform and transformation of state-owned enterprises and central enterprises cannot be realized only by the courage and efforts of their own. This can only be realized by the national will as well as the central government's decision-making and actions.

For core industries or those related to national security, their majority ownership should be controlled by the leading state-owned enterprises and central enterprises, i.e. controlled by the state. At the same time, the law of market economy should be followed. Finally, the mixed ownership mechanism will be formed in which the state-owned holding is the main body while the private and social capital coexist for a mutual development. In addition, the corporate governance structure such as professional managers should be introduced. Modern enterprise system could be employed to maintain the constant appreciation of state-owned assets and to satisfy the rights and interests of shareholders.

Fifth, the national education, science and technology systems should be adjusted and reformed. In the future, the essence of competition between nations, economies, industries and enterprises is a high degree of competitions on human resources. This requires a fundamental reform of the education system. When educating and training "special talents" for each industry or field, more "inter-disciplinary talents" with the knowledge of multi-format should be cultivated in the future.

The core of competition between nations, economies, industries and enterprises is more focused on the science and technology innovation in the future. It is necessary to carry out a fundamental scientific and technological reform. It is worth noting that technological development will lead to the innovations on not only production tools and productivity but also production relations, production mode, industrial development model, systems and mechanism. The national funding on scientific and technological research should place more focuses on the innovations of basic science and core technologies, and on the talent cultivation and team building. At the same time, the innovation activities with clear objectives of industries should be mainly implemented by enterprises themselves. Those key technological innovation activities of industries and enterprises that need the support of national policies should be with a view of the national strategic objects and tasks.

All in all, the support and in-depth application of software, Internet, 5G, big data and AI should be taken into account for everything in the future. A high degree of integration and unity is required between "offline" business and "online" management and service.

In the future, once such solid foundations have been laid, whether manufacturing enterprises are integrated with the software company and Internet company to implement "+ Internet" model, i.e. letting software company and Internet company "be married" to manufacturing enterprises, or the software company and Internet company are integrated with the manufacturing enterprise to implement "Internet + " model, i.e. letting manufacturing enterprise "be married" to software company and Internet company, both approaches will facilitate China's core and leading enterprises to develop into the most influential and competitive industrial Internet enterprises at the global scale. At this moment, it is easier to deal with whatever restructuring and reform of economy and industries. There will be a lot of potential and margin for the restructuring and reform of China's economy and industries in the future. In terms of the industrial format, it is the competitions between enterprises today. On the contrary, it will be the competitions between industrial ecosystems in the future where the Internet and industrial Internet companies are the core.

Furthermore, the structure, foundation and capabilities of China's manufacturing industry have once been adjusted and fine-tuned. With such solid foundation, the proportion of the service industry can naturally be greatly enhanced. This will allow the service industry to sell China's own products and service as much as possible.

Meanwhile, China's economic structure will be significantly extended in a large scale to realize a sustainable and healthy growth. China's economy and industries will be bigger and stronger, and the influence will be greater at the global scale. In this process, China's financial investment industry, which develops along with the manufacturing industry and the service industry, will also be promoted and achieve healthy development. The financial industry will therewith replace the manufacturing industry to play the dominant role of China's economy. In addition, its positioning will be increasingly clearer, enabling China to transform into a developed country. At the same time, the "concept" of Internet financial service industry with high efficiency will emerge. Based on big data and AI advantages, the Internet financial service industry (e.g. Alipay, WeChat and MI payment) will gradually become the reality of China's economic main body. Such Internet financial service industry will dominate China's economy in the future.

No one can influence and stop the great ambitions of China's economic reform. The Internet and the "e^3" economic development theory will certainly become the power to transform the Chinese economy.

4 Dealing with Significant Structural Issues Associated with the Financial Industry and Stock Market

China's financial industry is also a product of the planned economy. The Chinese banks and investment companies are derived from the state, which are lack of consciousness on competition and service. Their structural issues include: large and monopoly, large and comprehensive rather than strong and refined. Such structural issues of financial industry are similar to those of manufacturing industry. These will not be repeated in the following sections.

Compared with the United States and other western developed countries, it can be observed that China's financial industry lacks a century-long natural development process, i.e. pregnant, birth, growth and expansion. In the development process of financial industry of the United States and other western developed countries, through multiple rounds of fierce competitions, some of banks and investment companies were out of business while others survived and gained development. Since their birth, the Chinese banks and investment companies have been organized and protected by the state, without experiencing fierce competitions. Today, the major business model of these Chinese banks and investment companies is still providing loans and making investments on the basis of deposits or fund raising. Their income is mainly the interest margin, and the business model is very simple. In other words, such simple business model and business structure can no longer meet the needs of complex economic and industrial developments in China. This must be adjusted and reformed.

A further analysis is showed that, compared with the United States and other western developed countries, these China's banks and investment companies also lack such a century-long development process. Banks and investment companies of the United States and western developed countries have been working and collaborating together with the manufacturing industries and enterprises all the way. They have experienced success and failure, shared mutual developments, and finally formed natural cooperation links and mechanisms. On the contrary, China's banks and investment companies are basically nation-wide and zonal, each of which provides service to all industries and circles. Dislike the United States, China generally does not have vertical and business-specialized or industry-oriented banks and investment companies. This means that in China, the division between banks and manufacturing enterprises is distinctive. The relationship between financial industry and manufacturing industry is simply loan or funding, rather than a natural long-term business collaborative relationship. In other words, there is lack of deep business integration between these two industries to form a format of symbiosis and coexistence. There is no synergy pattern of $1 + 1 > 2$. Such simple business mode and simple structure cannot be sustained. Such structural issue will significantly influence the requirements of economic development, which should be adjusted and reformed.

The U.S. development process of financial investment industry and the changes to their business model have been discussed in the previous chapters. This deserves a serious consideration of China's financial industry as a useful reference.

It seems that as the main body of national economy, China's large state-owned banks and investment companies should consider to introduce a large amount of social capital; to make most of market economy's competition mechanism, and to strengthen their competition and service awareness. The deep integration with the Internet, 5G, software, big data, AI and other information technologies should also be strengthened to deal with the short slab of the informationization, in order for these state-owned banks and investment companies to provide strategic supports for the national core and leading industries.

Similarly, the large number of local state-owned banks and investment companies should consider to transform the highly marketization and socialized shareholding institutions, gradually becoming vertical, business-specialization or industry-orientation. Meanwhile, the in-depth cooperation with the software and Internet companies should be promoted. Finally such structure and pattern will be formed for the socialized division of labor with high level of rationality. They will complement and support each other with those large state-owned banks and investment companies.

At the same time, China's Internet and industrial Internet companies will definitely become a new financial format of the Chinese financial industry in the near future. China should conform to the trend of Internet economy and provide those Internet and industrial Internet companies with more operational space. The deep integration, win-win cooperation and sustainable development between the Internet/industrial Internet companies and the traditional financial banks/investment companies should be encouraged.

As a consequence, such system and structure will be finally formed in which the three main forms of the Chinese financial industry coexist and co-develop in China.

Let's take a look at the stock market. The opening-up of the Chinese stock market commenced in the 1990s. At that time, due to the severe shortage of funds, a large number of state-owned enterprises took the lead in stock market. Most of those enterprises belong to the traditional "asset-heavy" manufacturing and assembly industry with poor condition of assets, low level of performance and management, and high consumption of energy and resources. Apparently, it is understandable that these enterprises were publicly listed at that moment. However, from the perspective of international industrial pattern, a large number of those publicly listed state-owned enterprises are providing low to middle manufacturing and service with low added-value. This inevitably results in a hidden danger for the healthy and sustainable development of the Chinese stock market. Later, due to the influences of inertia thinking under the planned economy and possible temptations of some interest groups, more these sorts of enterprises with low capacity and efficiency continued to be publicly listed, forming the main body of the Chinese stock market. This finally resulted in structural issues associated with the Chinese economy.

Table A.14 of Appendix A shows the top 50 publicly listed companies in China in 2014. The similar results are arrived from the analysis of Table A.13. The top 50 publicly listed companies in China are mainly traditional banks, insurance companies, securities companies, traditional "asset-heavy" manufacturing enterprises and the real estate companies. This is a fundamental fact of current Chinese economy and industries. China extremely lacks the high-end manufacturing enterprises, software companies, Internet companies, "asset-light" industrial Internet companies, chip companies and Internet service banking. According to this fact, China's economy and industries are following a traditional development mode. There is a massive gap between the current mode and the new Internet economic development mode. Thus, we must work hard to catch up.

In United States and other western developed countries, the stock market is the financing platform for enterprises and the investment platform for investors. The stock market investment is generally long-term oriented. However, most of publicly listed enterprises in China are "asset-heavy" and are lack of excellent and high-quality assets. In addition, some publicly listed enterprises chronically do not pay their dividends to investors. The information of their management and operation is also opaque and asymmetric. foreign and domestic investment institutions and investors are not optimistic of these enterprises, resulting in a tendency for short-term investments or transactions. Usually, the investment institutions and investors will run away after earning some money, leading to frequent fluctuations of the stock market. Eventually this will result in an unique short-term stock trading platform or a so-called "betting platform" of the Chinese stock market, rather than a long-term investment platform. The market falls into such an embarrassing situation of "speculation" or "gambling". Such ups and downs of the stock market leads to the instability of enterprises, finance and economy, and even the society. Indeed, this is detrimental to the individuals, enterprises, industries and even the state.

Today, in the era of Internet economy, the international capital circle has reached a consensus. They have negative views of the manufacturing enterprises which adopt the traditional "asset-heavy" industrial development model. On the contrary, they are favour of those "asset-light" Internet companies and industrial Internet companies.

Under such circumstance, we should not continue to have a large number of publicly listed enterprises that adopt traditional "asset-heavy" industrial development model with low-quality and low-efficiency. This is because those enterprises have not yet achieved the industrialization, informationization and high-end manufacturing. Otherwise, the Chinese stock market will be dealing with an increasing number of serious problems. These will become the structural issues that impede the national economic development in China.

So, publicly listed enterprises in China must devote themselves to establish their own industrial technology systems and industrial ecological chains by achieving the industrialization, informationization, Internet, big data, AI and high-end manufacturing. With such capacities, they can be transformed into the "asset-light" Internet and industrial Internet companies. If so, it can be expected that the Chinese stock market will become an ideal financing platform and investment platform for both domestic and foreign investment institutions and investors. As a result, the Chinese stock market will provide strong support to the healthy development of national economy as well as promoting the social stability.

5 The Informationization Must Be Achieved at All Costs

In China, a large number of enterprises and their managers generally regard the software or informationization as a support tool for their management and business, which is auxiliary. They believe that the software, more or less, has a little impact on the development of enterprises. This has resulted in a misunderstanding of the strategic significance of software and information technology.

In fact, software, network and data have become the strategic assets of national economy, industries and enterprises, playing a crucial role in productivity and production relations. Today, high-end product's design has to be supported and implemented by software, network, and data. Internal and external management and control must be supported and implemented through software, networks, and data. As a result, point-to-point and end-to-end management can be achieved that is featured with highly flattening. This will significantly improve the management efficiency and reduce the costs. Meanwhile, the allocation of manufacturing resources must be accurately and efficiently completed by the software, network and data. The life-cycle service and personalized service of products must be realized through the software, network and big data. The decision making and strategy of enterprises must be supported and realized by the big data and AI. The software, network, data, AI and informationization services have become ubiquitous and all-encompassing. Indeed, it is crucial for an enterprise to make most of the software, network, big data and informationization. As a result, its corporate strategy, business model, resource management, business and process, product design and services can be achieved in high efficiency. In addition, enterprises must have the support of network, software and big data to transform into high-end manufacturing. Today, enterprises cannot operate and survive without the network, software, data and informationization.

From the deeper perspective, the support of network, software, data, AI and informationization is essential. Without such supports, the leading enterprise cannot establish its self-centered industrial technology system and industrial ecological chain, configure the superior resources, participate in the international competition, and be favored by the capital circles and stock market. Otherwise, they will not be able to transform to the "asset-light" industrial Internet development model and to achieve a shift from the product-oriented to the service-oriented. Therefore, the network, 5G, software, big data, AI and informationization are the thresholds for enterprises to realize the new industrial Internet development model.

I have visited hundreds of Chinese central enterprises, state-owned enterprises and private enterprises in various fields (e.g. household appliances, automobile, shipbuilding, aeronautical manufacturing, machine tools, construction machinery, locomotives, computer, communication equipment, energy and power). The results show that most of these enterprises have engaged in certain level of informationization and software development that focus on the four main links of enterprises (i.e. designing, manufacturing, management and service). They work very hard, and have made some progresses. However, it is necessary to discuss a universal phenomenon or fact. Those enterprises have heavily invested in developing the network, software, data and informationization. However, their bosses and managers are generally not satisfied. This is due to a large number of "isolated information island". Those independently developed software cannot be organically integrated so as to work together, e.g. sharing of information or big data. The faster and better the development of network, software and informationization, the more serious the phenomenon of "isolated information island". As a consequence, their bosses and managers at all levels cannot obtain all valid data needed for the operation, management and service of enterprise in a real-time and accurate manner. The data cannot be analyzed effectively for enterprises to make quick decisions and responses, which results in low efficiency. Therefore, it presents a significant challenge for enterprises to solve such serious "isolated information island" issues in order to achieve the informationization process and big data.

An in-depth analysis shows that each enterprise has made plans on its informationization, most of which however are in papers and are not operational. In fact, most of Chinese leading enterprises have not established a set of industrial standards and their digital standards based on their own businesses, processes, management and services. Similarly, they have not developed a set of standards that are used to define their industrial technology system and industrial ecology in a fine-grained manner. China has promulgated the national and some industrial standards. However, all these standards are coarse-grained, which cannot meet the actual needs for the development and operation of informationization and software. In many cases, these standards can only be used as a reference. As a result, the standards for product design used by enterprises will inevitably be those of design software vendors. The management software follow the standards of those management software vendors. The software related to processing, manufacturing and production process follow the standards of those related software vendors. The service software follow the standards of those service software vendors. Therefore, it is not hard to imagine the difficulties to mix together those sorts of software with respective different standards. It is surely not possible to achieve the effective operation, management, service and analysis of all data within an enterprise or an industrial ecosystem in a real-time and accurate manner. This prevents more efficient and accurate decision-making through the analysis of big data.

Therefore, to achieve the informationization process, the Chinese manufacturing enterprise must make every effort to establish a set of industrial standards and digital standards based on their own businesses, processes, management and services as its first priority work. Similarly, they should devote themselves to develop a set of standards that can define their industrial technology systems and industrial ecosystems. Accomplishment of digitization is subject to standardization. Accomplishment of network orientation is subject to digitization. Accomplishment of big data is subject to network orientation. Accomplishment of intellectualization is subject to big data. Based on these capacities, China's leading manufacturing enterprises can achieve the "asset-light" to transform into the new industrial Internet development model. It is a long and hard process to climb up such a "high mountain". Indeed, this is a new "long march" that calls for efforts from every single Chinese entrepreneur.

In China, a large number of leading enterprises have not completed the critical process of informationization. This presents significant challenges to achieve the modernization and to develop the Internet economy. If those issues cannot be solved in a timely manner, the informationization will become a new contradiction in the Chinese economic development. Enormous efforts are required to achieve the informationization process. The state, industries and enterprises must change their attitude, attach great importance and make great efforts to promote the informationization.

All in all, the time is now. Enormous pressures are presented from United States, Germany and other western developed countries that quickly move toward the Internet economy. Therefore, it is not appropriate to simply rely on enterprises themselves to solve those structural issues and contradictions that exist in China's manufacturing industry, financial industry, stock market and the informationization. It is imperative to implement reforms at the state and industry level. If these critical issues and contradictions cannot be solved in a timely manner, the Chinese enterprises will not be able to focus on the transformation and upgrading to develop the Internet economy. Similarly, they will not be able to actively deal with challenges that are derived from changes to the global economy. As a consequence, the Chinese manufacturing industry, financial industry and economy are likely to fall into such a trap in which industries will continue to be positioned as "asset-heavy" in the international economy (e.g. low-end processing, manufacturing and low-value services). Meanwhile, precious natural resources will continue to be consumed, and a huge environmental cost have to be paid. This will lead to an unsustainable development of the economy and society in China.

The duty is extremely heavy, and we must turn the pressure into the power. We have to deal with significant challenges associated with reforms of the Chinese economic and industrial structure. There are plenty of difficulties during the transformation and upgrading of China's industries and enterprises, however also with full of hopes.

Chapter 9
There Is a Great Journey Ahead to Initiate a New "Long March" of China's Internet Economy

The Internet and industrial Internet industries in the United States are developing aggressively, which is followed by Germany Industry 4.0 very closely. China's traditional industries follow the "+ Internet" model and Chinese leading manufacturing enterprises are transforming and upgrading in accordance with the "4S" development mode. Presented by significant challenges, these leading manufacturing enterprises have to find ways through. Meanwhile, China's emerging Internet companies follow the "Internet +" model, which will continually challenge the traditional manufacturing, financial and service industries.

1 Following the "+ Internet" Model to Sweep All Obstacles to the Transformation of Traditional Industries

The "asset-heavy" state-owned manufacturing industry forms the main body of the current Chinese economy. China's traditional industries and their leading enterprises should focus on the high-end manufacturing and the industrial Internet, practice the "e^3" economic theory, and adopt the "+ Internet" model. This conforms to the new "asset-light" Internet industrial development model, adapts to the new changes to the global economic pattern, and promotes the transformation and upgrading of traditional manufacturing and financial industries. This will be a great journey where significant challenges exist in a new "long march" of Internet economy. We should devote ourselves to sweep all obstacles.

There are four steps for the transformation and upgrading of China's traditional manufacturing and financial industries. The first step is to realize and complete the industrialization process as early as possible.

The market economy in China was not initiated until the economic reform and opening up in 1978. It is a very short period time, which is even shorter for most of Chinese enterprises. Therefore, the first step for the transformation and upgrading

J. Yuchi, *Value Proposition*,
https://doi.org/10.1007/978-981-15-5025-6_9

of China's manufacturing industry is to realize and complete the basic process of industrialization as soon as possible.

Under the market competition environment, China's leading enterprises in various industries should follow the international market competition cycles and be fully integrated into the international industrial systems. As a result, the rational socialized division of labor can be realized at a high rationalization level. Similarly, a transformation should be gradually achieved from the imitation to the creation. There must be a classic product for each of villages and towns in China while every product should be specialized and industrialized with high value. The SMEs are encouraged to be more specialization. Finally, a pattern is formed under which they are fused into the different industrial ecosystems of those leading enterprises. Such industrial ecosystem is featured with the labor division, mutual cooperation, interdependence, mutual support, promoting competition, survival of the fittest and orderly development.

China's financial industry should turn to be oriented to the industries and the sectors. Majority of banks and investment companies need to follow the industry/sector-oriented development. They should aim to achieve a transition from the current zonal and regional "blocks" structure to the industry/sector-oriented, specialized and longitudinal structure. Under the market competition, the financial industry as well as its banks and investment companies should realize the socialized division of labor at a high rationalization level so that a pattern can be formed under which they are deeply fused into those different industrial ecosystems of the leading enterprises. Consequently, the deeper integration and mutual support are established between the manufacturing industry and the financial industry. Finally, a new industrial ecology is formed.

The second step for the transformation and upgrading of China's traditional manufacturing and financial industries is from the industrialization to the informationization, realizing and completing the key process of the informationization.

When the industrialization is realized, China's manufacturing industry should achieve the transformation and upgrading into the informationization as soon as possible.

China's leading manufacturing enterprises in various industries should constantly implement the "software-orientation" and informationization. They should work closely with the core R&D institutions to establish fine-grained industrial standards and digital standards so that their respective industries can be defined. This is the priority strategy for the nation, industries and enterprises. Based on such foundations and by cooperating with the core software companies, manufacturing enterprises should develop and apply a large number of software on those four critical links (i.e. product design, resource management, processing and manufacturing, and service). This helps to achieve their standardization manufacturing, digital manufacturing, network manufacturing and intelligent manufacturing. This is another priority strategy for the nation, industries and enterprises. Consequently, these leading enterprises are able to build self-centered industrial technology systems as well as self-centered industrial supply chains, service chains and ecological chains. On this basis, they should lead and support a large number of SMEs in the whole society around the world to join their industrial ecosystems. Finally, the key process of informationization is completed. Thus, these leading enterprises and majority of specialized SMEs could cooperate more closely in the industrial ecology.

In other words, by realizing the informationization process, China's leading manufacturing enterprise should have the capacities to control over the industrial standards, industrial technology system, industrial ecology and the allocation of superior industrial resources at the global scale. As a consequence, the Chinese leading manufacturing enterprise will be transformed to a composite entity that covers various businesses and industries such as: core technology R&D, industrial standard, service, software, Internet, big data-AI-consulting service and business management.

It is worth noting that there is a high-cost associated with the realization of informationization in the manufacturing industry. For the majority of SMEs, such process is very expensive and therefore not affordable. Under such circumstance, the informationization cost of SMEs in fact forms the cost element of the manufacturing industry as well as the nation. Therefore, the best solution is that the state and industries intensively support the informationization of those leading enterprises. Then, the majority of SMEs can acquire or be allocated all sorts of software and services required to realize informationization. This can be achieved by joining those industrial ecosystems established by the leading enterprises.

There is a massive gap in the financial informationization between China and developed countries such as the United States. Significant challenges exist for the realization of financial informationization in China. The information management of those Chinese banks and investment companies should be realized. Meanwhile, they should also achieve a transformation from product-oriented to the business-oriented and service-oriented. Subsequently, they should focus on certain industries or sectors, strengthen in-depth cooperation with both the software companies and the manufacturing enterprises, and vigorously develop all sorts of the management and business software for their customers. This means that the business model will be transformed. Those specialized banks and investment companies are able to provide the professional software (i.e. management and business services) and networking tools to the customers so that they can achieve the direct and real-time connections with their customers. By analyzing related big data and data transparency, those banks and investment companies can provide valuable professional consulting services to their customers. Meanwhile, they can reduce the risk of their loans and investments; and enhance and improve their efficiency to the largest degree. As a result, the risk control and sustainable development can ultimately be achieved for the Chinese financial industry.

The third step for the transformation and upgrading of China's traditional manufacturing and financial industries is to March into the high-end manufacturing industry.

With strong support of the industrialization and informationization, China's leading manufacturing enterprises should stride into the third step. They should be transformed into the high-end manufacturing enterprises.

High-end manufacturing is lucrative but highly competitive. Therefore, it is crucial that enterprises have a global vision, a huge amount of capital inputs, and a capacity to allocate the global superior resources. China's leading enterprises should continuously pursuit innovations and make full most of the large domestic market demand. Similarly, they should aim to attract massive investments from the financial industry and stock market, and allocate the best resources around the world. As a consequence, they can seize the control of core industrial technologies, core industrial standards, intellectual property, and core industrial values.

The high-end manufacturing industry is highly profitable and great potential. Therefore, in future, the high-end manufacturing industry will for sure deal with challenges from the Chinese Internet industry. However, competition inspires advancement. Over time, it can be foreseen that China's high-end manufacturing industry will deliver good news as expected.

In this process, with the support of the informationization and industrialization, China's financial industry and social capitals must keep the full speed; and establish closest cooperation with the manufacturing industry and Internet companies. They should provide full supports; and dare to invest a large amount of capitals to make greater achievements. A powerful high-end manufacturing will make both manufacturing industry and financial industry stronger. In future, the financial investment industry and stock market must form an integral part of high-end manufacturing in China.

The fourth step for the transformation and upgrading of China's traditional manufacturing and financial industries is to adopt the "+ Internet" development model and March into the industrial Internet industry.

China's manufacturing industry will inevitably be upgraded again with the solid foundation of the industrialization, informationization and high-end manufacturing. By adopting the "+ Internet" model, China's manufacturing industry will March into the industrial Internet industry, accompanied by the financial industry and social capitals. At that moment, the manufacturing industries and enterprises will be transitioned from traditional "asset-heavy" development mode to the new "asset-light" industrial Internet development mode. In the new development mode, the operation of Chinese leading enterprises places more focuses on the core values of industrial ecology. This helps to achieve a transition of enterprises from product-oriented to the service-oriented. China's leading manufacturing enterprises will be transformed and upgraded into such composite entities that cover a variety of industries and businesses such as core technology manufacturing, industrial standards, industrial software, Internet service, big data-AI-consulting services, financial services, business management and so forth. Through continuous innovation, these enterprises can control the core industrial technologies, industrial standards, industrial technology systems, industrial supply chains, industrial service chains and industrial ecological chains. As a consequence, these enterprises can occupy the top of their own industrial value chains, with an aim to pursue the core values of multi-industry and mult-business. They are the main driving force of China's economy and industries.

In this process, the Chinese manufacturing industry will be deeply integrated with its associated service industry, software industry, etc. Meanwhile, most of the traditional banks and investment companies will be transformed to provide services to some certain manufacturing industries and sectors. They will oversee and control their related manufacturing industries. In the future, the boundary between the financial industry and the Internet/industrial Internet industries will become increasingly blurred. In addition, the traditional banks and investment companies will be increasingly inclined to provide their loans and investments through the Internet and industrial Internet financial service companies. In fact, it is inevitable that the Internet/industrial Internet companies will be transformed to new banks and financial investment companies. In other words, the Internet/industrial Internet companies will play a critical role in the financial industry. Therefore, it warrants further observation and discussion whether a considerable proportion of traditional banks and investment companies will independently exist by themselves in the future.

Following aforementioned four steps, four transformations and upgrading will be achieved. As a result, China's leading manufacturing enterprises, or those new industrial Internet companies that are industry-oriented, would March steadily towards the Internet economy. At this moment, China's industrial Internet companies will be fully involved in the international competition. Similarly, they will win the respect of their opponents. Those opponents frequently made short-term investments in Chinese stock market will become the long-term investment partners from now on. At this moment, China's manufacturing industry and the financial industry will "be liberated". Indeed, it can be proudly stated that China is manufacturing power, financial power, and economic power. At the same time, China will fully embrace the era of Internet economy.

The Chinese enterprises, industries and economy will go through such four main steps for their transformation and upgrading, i.e. industrialization (first step), informationization (second step), high-end manufacturing (third step) and industrial Internet (fourth step). I would like hereby to define such a long march as the Chinese "4S" industrial development mode. It should be noted that these four steps can be implemented one after the other or can be better implemented in parallel. This will speed up the process to keep up with the global Internet economy.

Following the "4S" industrial development mode, at present, publicly listed enterprises in China's stock market should mainly consist of those with the completion of the first two steps, namely the realization of industrialization and informationization. In the future, China's publicly listed companies in stock market should mainly consist of those with the completion of the last two steps, namely the realization of high-end manufacturing and industrial Internet. These Chinese leading enterprises play a major role in the international competition.

In the Internet economy era, China's traditional manufacturing, service, financial industry and stock market will be completely transformed. The maximization of industrial values will be achieved with developed financial investment industries. Investment risks will be controllable. The pressure on the environmental pollution discharge will be manageable. With enhanced discourse rights on economy and industries, they will gain strong international competitiveness.

2 Following the "Internet +" Model, the Internet Industry and Its Companies Will Smash Opponents

The rapid development of Internet and the industrial Internet in United States will introduce profound influence on China's Internet industry, software industry, service industry, financial investment industry and manufacturing industry. Challenges and opportunities co-exist.

"Internet +" model is completely transforming the Chinese traditional service industry today.

Service industry has existed for thousands of years. Today, Internet companies are constantly transforming the traditional service industry. These Internet companies include: Google and Baidu (search engine, digital library, smart city and smart earth), Apple and MI (digital household appliance and Apps), Amazon, Alibaba and JD (e-business), Tencent (Wechat) and Facebook, Uber and DiDi, SF Express, etc. Wherever it goes, the traditional service industry is collapsed successively. The service industry is experiencing the industrial revolution. Driven by information technologies (e.g. Internet, 4G, 5G, cloud-computing, software, big data and AI), a new type of service industry emerges that is created by Internet companies. In this new industry format, all services can be achieved or supported through the Internet and mobile terminals. All services can be provided with higher efficiency, shorter response time, lower cost and better service experience.

The United States and other western countries are classified as developed countries because of their developed service industries, comprehensive service systems and convenient services. Perhaps their original service systems and services are well received, the transformation of those developed countries to the Internet service seems quite slow today. As a typical example, the U.S. Internet operators (e.g. Amazon and Yahoo) are developed much slowly than China's Alibaba and JD. Using the service industry as the starting point, China's Internet industry is promoting the reform of the traditional service industry. Its development is faster than that of the United States and other western developed countries. The great-leap-forward development of China's service industry can be achieved through the "Internet +" model, which is worth pondering.

The Chinese service industry is now well supported and operated by Internet companies such as Alibaba, Tencent, Baidu, JD, Sina, Sohu, DiDi, Ctrip, SF Express, 360 and so on. As a result, China's service industry can keep pace with the times as long as the governmental policies are in place and old rules are broken.

"Internet +" model will completely transform the traditional manufacturing industry in China, which will ensure this industry March into the high-end manufacturing industry.

China's traditional manufacturing industry adopts "+ Internet" model to enter the industrial Internet industry. By contrast, China's Internet industry employs the new "Internet +" and "asset-light" industrial Internet development model to enter the manufacturing industry and real economy. This is achieved by taking advantages of information technology such as Internet, 5G, software, cloud-computing, large data and AI. In particular, the advantages of financing are made most of. Upon the allocation of global and national superior manufacturing resources, the Internet industry will be also developed into the industrial Internet industry with more obvious advantages.

Today, China's manufacturing industry is featured with large scale and massive domestic market. Once entering manufacturing industry and becoming the leading industrial Internet enterprises, these Chinese Internet companies will likely be ranked the first in that sector globally in terms of size and potential. Therefore, China's Internet/industrial Internet companies will inevitably be recognized and sought after by the international and domestic capitals and stock markets. Its market value will be much more increased, and the financing capacity will be enormously enhanced. As a consequence, as long as some manufacturing sectors are profitable, China's Internet companies will inevitably adopt the "Internet +" model and the new "asset-light" industrial development model to enter the manufacturing industry. Their pace will be faster, with more open business ideas and more international visions. If they are able to integrate the superior manufacturing resources throughout the country and even the world, and invest large amounts of money to develop and control the industrial core technologies and the industrial standards, the success of "Internet + " model will be achieved sooner. As a result, the deep integration of the Internet industry and manufacturing industry can be achieved, i.e. "1 + 1 > 2". Therefore, the "Internet +" model will have greater impacts on China's industrial development in the future. Similarly, the Internet industry will completely transform the traditional manufacturing industry as well as the traditional industrial development model in China.

As mentioned in Chaps. 6 and 8, MI used smart phones as the starting point and shares product designing with its users and fans through the Internet. Its processing, manufacturing, assembly and testing processes are all outsourced. This is controlled through MI's industrial standards, industrial technical system and industrial ecological chain. Through the network, software, big data and AI, the control of its internal and external management, and service are achieved to minimize costs of manufacturing and management. Similarly, it helps to maximize management efficiency, to improve the market responsiveness, to minimize the product price, and to maximize shareholders' profits. Although it was only established for ten years, MI has ranked the top level in China and even in the global market in terms of sales of smart phones, posing a challenge to Apple and Samsung. The most important thing is that MI adopts the "Internet +" model, which is "de-industrialized" and "asset-light". Similarly, MI has been recognized and sought after by the domestic and foreign capitals. Its market value has been estimated as high as $100 billion. With such a strong capital strength, MI is completely transforming China's smart phone industry and market. At the same time, MI is looking into the digital home appliances and traditional white goods market. It is likely that MI will completely transform the industry and market of digital home appliances and traditional white goods in China and overseas in the future. Indeed, as the first industrial Internet company in China, MI's great success can serve as useful reference for Chinese entrepreneurs, industry circle, financial investment circle, economy circle and the governments.

In fact, fake goods in Alibaba and their regulations have been frequently discussed on the Internet. We can guess and expect that in the future, Alibaba will certainly adopt the "Internet +" industrial development model to advance into the real economy and the relevant manufacturing domains. This means that in the future, a growing number of Alibaba-branded genuine products will be launched in the market. Vast number of consumers will be assured to make a buying decision with higher price-quality ratio and better consumer experience. Today, a growing number of Chinese enterprises and individuals begin to use Tencent's WeChat, which provides the most convenient functions of social services. Meanwhile, Tencent is developing the corporate information software. It can be expected that Tencent will gradually March into the real economy and manufacturing industry, even the industrial Internet industry, with the corporate informationization, OA software and service as starting points.

The Internet industry will surely enter into the high-end manufacturing industry. The development of high-end manufacturing industry requires huge amount of capital inputs. Let's take a look at the market value of several Chinese Internet companies. Publicly listed in United States, Alibaba has the market value of more than $490 billion. Publicly listed in Hong Kong, Tencent recently has the market value of more than $530 billion. MI has been estimated with a market value of more than $100 billion. It can be observed that these Internet companies are good at financing. They are extremely wealthy. If the high-end manufacturing is more profitable, China's Internet companies will for sure make heavy investment, and adopt the "Internet +" model to enter those industries. It can be expected that more high-end manufacturing products will be provided by China in the near future.

The Internet companies will develop the high-end manufacturing from another perspective. Those Chinese Internet companies could rely on their sufficient financial advantages and integrate global resources (rather than domestic resources) required by high-end manufacturing. If so, China's high-end manufacturing will have the opportunity to succeed. The Chinese Internet companies are born with an international vision, which is natural. From the perspective of Internet industry and their operators, it is a parallel world. In other words, the earth is simply a village. Why the high-end manufacturing industry has to be developed through the completely independent Chinese innovation? In fact, the key lies in the party that can master and control those needed core technologies, core industrial standards and the core values of high-end manufacturing.

"Internet +" model will present strong vitality and lead China into the Internet economy.

Since its birth, China's Internet industry has gained development in the context of market economy and international competition. China's Internet industry is completely transforming the traditional service industry. Meanwhile, by adopting the "Internet +" model, it is moving downstream towards the real economy to transform the traditional manufacturing industry. Similarly, China's Internet industry is also relying on its huge number of users and fans to move upstream towards the financial investment industry. As a result, its financialization and service are widely accepted and achieved, and a new format of financial industry is created now in China. In the future, it will for sure subvert the traditional financial industry and show a new pathway for the transformation and upgrading of China's traditional financial industry. At this moment, the Chinese traditional financial industry will have no choice but to be open-minded to integrate with the Internet industry and industrial Internet industry so that its "regeneration" can be achieved. At this critical moment, the state must no longer hesitate to formulate strong policies to facilitate the development of the Internet industry and Internet financial industry.

China's Internet industry is consolidating the resources of software industry. This means that the software industry will become a part of the Internet industry and industrial Internet industry in the future. In other words, the software industry may fade away and be merged into the Internet industry. In fact, the demise of an old software industry indicates the birth of a new software industry. In this process, the Internet industry, the software industry and the financial industry will be integrated each other. Similarly, they will be seamlessly integrated with the manufacturing industry and its associated services industry. A great platform service economy based on the Internet, 5G, "Cloud", big data and AI is coming to each of us. The Internet industry will present a strong vitality and completely transform the Chinese traditional service industry, manufacturing industry, financial industry, even software industry and so on. It will become a new and leading driving force for China to advance into the era of Internet economy.

In the era of Internet economy, the telecommunications industry will become the foundation and key link for the great development of Internet industry and industrial Internet industry. Thus, the telecommunications industry will become a "special" industry. The telecommunications industry must provide services with larger bandwidth, fast-speed, wireless and cost-effectiveness. If the telecommunications industry wins, China's Internet industry and industrial Internet industry will lose, and China's Internet economy will fail; and vice versa. In the future, the service cost of telecommunications industry will become the cost of Internet economy and the cost of China's future economy. Therefore, the Chinese telecommunications industry should prioritize the national benefits by providing strong support for China to advance into the Internet economy.

There is an interesting phenomenon which has profound impact on China and beyond. In the world, no one would anticipate that so many Chinese people like to use the Internet. This might be attributed to the pursuit of advanced productivity in the Chinese culture. Perhaps, the services originally provided in western developed countries are so good that has affected their transition from the traditional service to the new Internet service. This interesting phenomenon presents significant opportunities for the Chinese Internet industry and industrial Internet industry to achieve great development.

It can be expected that China's Internet industry will gain rapid development like "splitting the bamboo" by taking the "Internet +" model. The Internet and industrial Internet present a historical opportunity for China to achieve the status of developed country. An era of Internet and industrial Internet is fast approaching.

Chapter 10
A New Blueprint Is Fast Approaching with the Spring Breeze

Internet economy, new theory, new structure, new mode, new pattern and new format

In the past, China has proposed the integration, and even the deep integration between industrialization and informationization. In the future, in the context of new economic theory (i.e. Internet economy), China's Internet industry will be highly integrated with all industries such as manufacturing, service, financial investment, chip and software.

The global economy is experiencing profound changes. As the main driving force of national economy, China's traditional manufacturing industry, financial investment industry and service industry will overcome significant challenges to gain growth by adopting the "+ Internet" model. As the emerging driving force of national economy, China's Internet industry is rapidly developing by adopting the "Internet +" model. In the future, these two forces, i.e. "+ Internet" and the "Internet +" will meet and join together in "asset-light" industrial Internet industry. At that moment, China has gradually transformed its economic structure and industrial development mode. Originally, it is simple and backward mode, e.g. "heavy state-owned", "heavy manufacturing" and "heavy-asset". In the new mode, i.e. "asset-light" industrial Internet mode, the corporate ownership is dominated by the mixed economy; the state-owned corporates are mainly owned by leading enterprises of the national core industries; and the private and the state-owned corporates are mutually collaborated while domestic and international corporates can smoothly cooperate with each other.

As for the economy, the manufacturing industry serves as a solid foundation. Meanwhile, the service industry serves as the extension of manufacturing industry and economy. In addition, the Internet financial service industry plays a leading role and is integrated with the traditional financial industry. Furthermore, the Chinese leading enterprises mainly adopt the "asset-light" industrial Internet development model. They have established their industrial ecosystems where these leading enterprises can strongly support the majority of specialized SMEs each other with a high level of synergy. Finally, such new economic and industrial structure is formed. With

J. Yuchi, *Value Proposition*,
https://doi.org/10.1007/978-981-15-5025-6_10

a consideration of Chinese characteristics, the "e3" economic development theory will provide an essential guide for China to advance into the new Internet economy.

With the support of Internet, 5G, big data and AI, reforms will take place in production relations, production mode, industrial development model and economic base. As a result, a new superstructure of China will be developed. At that time, the quality of national economy will be improved, so as the capability to deal with economic risks. China's economy will be greatly strengthened so that China becomes a genuine economy power and financial power. China will have sufficient strength to lead the international convention on the reduction of atmospheric emissions as well as the world economic pattern. At that time, with strong innovation capacity and core competitiveness, China's leading enterprises will occupy the top of industry's core value chain at the global scale. As a result, their investment risks can be controlled, investment benefits will be maximized, and strong capabilities will be gained for the sustainable development.

In the Internet economy, China's economy will present new pattern and new format. In the future, China will witness 3-5 smart Internet integrated service providers (e.g. Alibaba, Tencent and Baidu) that are equivalent to those large-scale department stores of today. The competition among them will be extremely fierce. Similarly, their services will be all-encompassing, such as smart city, smart China and smart Earth, etc. Each of the smart Internet integrated service providers will be surrounded by an industrial ecosystem that consist of banks, financial investment and insurance companies, software companies, big data and AI consulting companies, a large number of service enterprises and massive people. Apparently, in the future, the main competition is among these industrial ecosystems.

At the same time, there will be tens of thousands of the industrial Internet companies like MI that look after various industrial sectors or businesses. They correspond to all sorts of franchise stores today. Each industrial Internet company is surrounded by its industrial ecosystem that consists of banks, financial investment and insurance companies, software companies, big data and AI consulting companies, the majority of specialized small to medium manufacturing enterprises, a large number of associated service enterprises and massive people. They will provide the industrial and professional services to associated production activities and the entire society.

These Industrial Internet companies may also cooperate with the smart Internet integrated service providers to form a new industrial format and a new economic structure. They work in cooperation with a due division of labor, while competing with each other. Apparently, such industrial ecosystems form the broader industrial supply chains, service chains and ecological chains, which will include more and more SMEs for their mutual and orderly development. This will make each village, town and city distinctive in terms of providing unique and competitive products. Indeed, this will lead to better quality of life and prosperous society.

In the future, everyone will love and rely on smart phones and intelligent mobile terminals. The data center will be everywhere. Computing and storage services will be used as convenient and low-cost as water and electricity. All services are rendered on the cloud, and the cloud services and 5G will be everywhere. Every household in urban and rural areas can surf the Internet with high bandwidth speed in a fast and cheap manner. No matter where you are, you can enjoy the warm-hearted Internet services as long as there is the Internet, 5G and smart phone in place. Everyone can benefit from the Internet, 5G, big data, AI and block-chain service. In short, all social and production activities in the future will be very convenient which are orderly undertaken in the Internet and industrial Internet. Everyone is happy enjoying the quality of life. In addition, the nation, industries and enterprises have gained powerful strength.

The new "Long March": grand and magnificent

More than 80 years ago, a great man, comrade Mao Zedong led 300,000 warriors of the Red Army to a 25,000-Chinese-mile "long march" from the famous Jinggang Mountain of China. Only 30,000 warriors survived and arrived at the Holy City of Yan'an after numbers of fights and overcoming all sorts of difficulties. Those 30,000 heroes had played a crucial role in the founding of new China in 1949. These heroes deserve respect.

Spring breeze blows a thousand layers of waves, and hundreds of boats compete by sails. Today, a grand new "long march" to the Internet economy will be initiated, i.e. restructuring the economy, reform of supply side, transforming the industrial development mode, transforming and upgrading of enterprises. This process is bound to be full of challenges. 30,000 Internet and industrial Internet companies may involve in this process. They have the courage to build a new industrial ecology and economic format through the continuous innovation. If that is the case, those 30,000 companies are the heroes of Internet economy in the future. They are the pillars of the national economy and backbones of the nation. Indeed, they present the hopes of China and Chinese people to gain powerful strength in this century.

Appendix
Summary of Tables

See Tables A.1, A.2, A.3, A.4, A.5, A.6, A.7, A.8, A.9, A.10, A.11, A.12, A.13 and A.14 .

Table A.1 Global acquisition of software companies by Germany's Siemens

Time of acquisition	Name of acquired company
2007	UGS (USA) (acquisition amount: \$3.5 billion)
2008	Innotec (German)
2009	Elan Software System (France)
2011	Active SA (Brazil)
2011	Vistagy Inc. (US)
2012	Senergy sistemas de Medicao S.A. (Brazil)
2012	LMS International (Belgium) (acquisition amount: 0.68 billion EUR)
2013	TESISPLM ware (German)
2014	CAMSTAR (USA)
2015	Kineo CAM (France)

Source from statistical information on Baidu

J. Yuchi, *Value Proposition*,
https://doi.org/10.1007/978-981-15-5025-6

Table A.2 Rankings of top 10 software vendors in the world ($100 million)

2013 Rankings	2012 Rankings	Software vendor	Software revenue in 2013	Software revenue in 2012	%YOY increase
1	1	Microsoft	657	620	6
2	3	Oracle	296	287	3.4
3	2	IBM	291	287	1.4
4	4	SAP	185	169	9.5
5	5	Symantec	64	64	0
6	6	EMC	56	54	4.9
7	7	HP	49	50	−2.7
8	9	VMware	48	42	14.1
9	8	CA	42	43	−2.6
10	12	Salesforce	38	29	33.3

Data source Gartner

Table A.3 Rankings of the top 20 Internet companies in the world in terms of market value (Unit $100 million)

Name	Country	Market value
Google	United States	3900
Facebook	United States	1939
Alibaba	China	1650
Amazon.com	United States	1496
Tencent	China	1476
Baidu	China	739
eBay	United States	633
The Priceline Group	United States	605
Yahoo!	United States	423
JD.com	China	395
Twitter	United States	300
Netflix	United States	275
Linkedln	United States	255
Yahoo!Japan	Japan	222
Naver	Korea	204
Rakuten	Japan	160
TripAdvisor	United States	136
Equinix	United States	114
NetEase	China	113
Vipshop	China	112

Source S&P Capital IQ, staff reports (Alibaba)

Table A.4 Rankings of top 25 IT enterprises in the world in terms of market value

Ranking	IT enterprise	Market value/$100 million	Stock price as of June 30, 2014	Stock price as of March 31, 2014
1	Apple	5,603.4	92.47	75.87
2	Google	3,891.1	575.28	556.97
3	Microsoft	3,444.6	41.44	40.45
4	Samsung electronics	2,163.8	1,322,000	1,343,000
5	IBM	1,834.8	180.2	190.25
6	Oracle	1,805.6	40.41	40.67
7	Facebook	1,749.6	67.29	60.24
8	Intel	1,538.2	30.7	25.42
9	Amazon	1,494.5	324.78	336.37
10	Tencent	1,424.8	118.2	107.66
11	Qualcomm	1,336.8	78.76	78.02
12	Cisco	1,273	24.66	22.06
13	TSMC	1,098.6	20.92	19.58
14	SAP	952.8	77	79.83
15	Tata consultancy services	789.9	2,420.75	2,108.82
16	Baidu	654.6	186.81	152.27
17	eBay	634.4	50.06	55.24
18	Priceline.com	630.8	1,203	1,191.89
19	HP	630.3	33.54	32.07
20	EMC	539.3	26.34	27.29
21	Texas instruments	515.8	47.49	46.56
22	Accenture	510.3	80.84	78.77
23	VMware	416.8	96.81	108.02
24	Asml	416	93.27	92.43
25	Ericsson	397.7	12.08	12.86

Source S&P Capital IQ, staff reports (Alibaba), The Wall Street Journal

Table A.5 Top 500 global enterprises in 2000 ranked by the Fortune Magazine (1–50)

Rankings	Name	Location of headquarters	Primary service	Operating revenue/$1 million
1	General Motors	United States	Automobile	176,558.00
2	Wal-Mart Stores	United States	Retail	166,809.00
3	Exxon Mobil	United States	Oil refining	163,881.00
4	Ford Motor	United States	Automobile	162,558.00
5	DaimlerChrysler	Germany	Automobile	159,986.00
6	Mitsui	Japan	Trading	118,555.00
7	Mitsubishi	Japan	Trading	117,766.00
8	Toyota Motor	Japan	Automobile	115,671.00
9	General Electric	United States	Electronic and electrical	111,630.00
10	Itochu	Japan	Trading	109,069.00
11	Royal Dutch/Shell Group	Netherlands/Britain	Oil refining	105,366.00
12	Sumitomo	Japan	Trading	95,701.60
13	Nippon Telegraph and Telephone	Japan	Telecom	93,591.70
14	Marubeni	Japan	Trading	91,807.40
15	AXA	France	Insurance	87,645.70
16	Intl. Business Machines	United States	Computer	87,548.00
17	BP Amoco	Britain	Oil refining	83,566.00
18	Citigroup	United States	Finance	82,005.00
19	Volkswagen	Germany	Automobile	80,072.70
20	Nippon Life Insurance	Japan	Insurance	78,515.10
21	Siemens	Germany	Electronic and electrical	75,337.00
22	Allianz	Germany	Insurance	74,178.20
23	Hitachi	Japan	Electronic and electrical	71,858.50
24	Matsushita Electric Industrial	Japan	Electronic and electrical	65,555.60
25	Nissho Iwai	Japan	Trading	65,393.20
26	U.S. Postal Service	United States	Mail parcels and freight transport	62,726.00
27	Ing Group	Netherlands	Insurance	62,492.40

(continued)

Table A.5 (continued)

Rankings	Name	Location of headquarters	Primary service	Operating revenue/$1 million
28	AT&T	United States	Telecom	62,391.00
29	Philip Morris	United States	Food, and tobacco	61,751.00
30	Sony	Japan	Electronics	60,052.70
31	Deutsche Bank	Germany	Bank	58,585.10
32	Boeing	United States	Aviation	57,993.00
33	Dai-ichi Life Insurance	Japan	Insurance	55,104.70
34	Honda Motor	Japan	Automobile	54,773.50
35	Assicurazioni Generali	Italy	Insurance	53,723.20
36	Nissan Motor	Japan	Automobile	53,679.90
37	E.ON (Veba and Viag)	Germany	Trading	52,227.70
38	Toshiba	Japan	Electronic and electrical	51,634.90
39	Bank of America Corp.	United States	Bank	51,392.00
40	Fiat	Italy	Automobile	51,331.70
41	Nestle	Switzerland	Food	49,694.10
42	SBC Communication	United States	Telecom	49,489.00
43	Credit Suisse	Switzerland	Bank	49,362.00
44	Hewlett-Packard	United States	Computer	48,253.00
45	Fujitsu	Japan	Computer	47,195.90
46	Metro	Germany	Retail	46,663.60
47	Sumitomo Life Insurance	Japan	Insurance	46,445.10
48	Tokyo Electric Power	Japan	Power supply	45,727.70
49	Kroger	United States	Retail	45,351.60
50	Total Fina Elf	France	Oil refining	44,990.30

Table A.6 Top 500 global enterprises in 2015 ranked by Fortune Magazine (1–50)

Rankings	Name	Operating revenue/$1 million	Profit/$1 million	Location of headquarters
1	Wal-Mart	485,651	16,363	United States
2	Sinopec Group	446,811	5,177	China
3	Royal Dutch Shell	431,344	14,874	Netherlands
4	China National Petroleum	428,620	16,359.5	China
5	Exxon Mobil	382,597	32,520	United States
6	BP	358,678	3,780	UK
7	State Grid	339,426.5	9,796.2	China
8	Volkswagen	268,566.6	14,571.9	Germany
9	Toyota Motor	247,702.9	19,766.9	Japan
10	Glencore	221,073	2,308	Switzerland
11	Total	212,018	4,244	France
12	Chevron	203,784	19,241	United States
13	Samsung Electronics	195,845.3	21,922.7	Korea
14	Berkshire Hathaway	194,673	19,872	United States
15	Apple	182,795	39,510	United States
16	Mckesson	181,241	1,476	United States
17	Daimler	172,279.1	9,235.3	Germany
18	Industrial and Commercial Bank of China	163,174.9	44,763.9	China
19	Exor Group	162,163	428.5	Italy
20	Axa	161,173.4	6,664.5	France
21	General Otors	155,929	3,949	United States
22	E.ON	151,460.5	4,191.8	Germany
23	Phillips 66	149,434	4,762	United States
24	General Electric	148,321	15,233	United States
25	Eni	147,175.8	1,712.6	Italy
26	Gazprom	144,408.5	4,124.4	Russia
27	Ford Motor	144,077	3,187	United States
28	Petrobras	143,657	7,367	Brazil
29	China Construction Bank	139,932.5	36,976.6	China
30	CVS Health	139,367	4,644	United States

(continued)

Table A.6 (continued)

Rankings	Name	Operating revenue/$1 million	Profit/$1 million	Location of headquarters
31	Hon Hai Precision Industry	139,039.4	4,307.8	China Taiwan
32	Allianz	136,846.2	8,252.3	Germany
33	AT&T	132,447	6,224	United States
34	Valero Energy	130,844	3,630	United States
35	Unitedhealth Group	130,474	5,619	United States
36	Agricultural Bank of China	130,047.7	29,126.4	China
37	China State Construction Engineering	129,887.1	2,079.1	China
38	Japan Post Holdings	129,686.7	4,390.1	Japan
39	PDVSA	128,439	7,386	Venezuela
40	Trafigura Beheer	127,612.6	1,039.7	Netherlands
41	Verizon Communications	127,079	9,625	United States
42	BNP Paribas	124,333	208.3	France
43	Lukoil	122,803	4,746	Russia
44	Honda Motor	121,221.5	4,633.4	Japan
45	Bank of China	120,946	27,525.1	China
46	Amerisourcebergen	119,569.1	276.5	United States
47	Pemex	119,238.7	19,929.4	Mexico
48	Assicurazioni Generali	118,871.7	2,215.3	Italy
49	SOCIéTé GéNéRALE	118,232.3	3,571	France
50	Fannie Mae	116,461	14,208	United States

Source from Chinese website of Fortune Magazine

Table A.7 Market value of Japanese home appliance companies and American Apple company

Company name	Market value/$100 million
Sony	338.81
Panasonic	167.24
Hitachi	339.2
Sharp	34
Apple	6,996.75

Sourcing from stock value as of November 17, 2015

Table A.8 Market value of general-purpose CPU chip providers and computer and server OEMs

Company name	Market value/$100 million
Intel	1,636
Dell	216
Lenovo	122
Inspur	48

Sourcing from stock value as of November 17, 2015

Table A.9 A&M cases by Internet companies in the last decade

Acquirer	Acquired company
Google	You Tube Video site
Google	Zagat US restaurant reviews website
Google	Quick office Mobile office software
Google	Waze Maps start-up company
Yahoo	Stamped Mobile recommended applications
Yahoo	On The Air video communications technology company
Yahoo	Snip.it Content integration platform
Yahoo	Go Poll Go Online polling company
Yahoo	Admovate Mobile advertising company
Yahoo	Social platform big data company Ztelic
Facebook	Be Lug Mass text information service tool
Facebook	Day Tum Personal visualization data tool
Facebook	Friend.ly Social Q&A site
Facebook	Instagram Picture sharing application
eBay	StumbleUpon website recommended service company
Microsoft	Bungie company
Microsoft	Skype
JD	QQ online shopping and paipai.com
Alibaba	Buying shares of Youku Tudou
Alibaba	Autonavi
Tencent	58.com
Tencent	Dianping.com
Baidu	Nuomi.com
Baidu	91.com
Alibaba	Sina Weibo
Tencent	Sogou
China Telecom	tv189.com
Qihoo	Ruichuang Network
Suning.cn	Red baby

Table A.10 Market value of China's telecom operators, telecom equipment suppliers and Internet service providers

Company name	Market value/$100 million
China Mobile	2,423
China Unicom	227
China Telecom	420
ZTE	122
Tencent	1,881

Sourced from stock value as of November 17, 2015

Table A.11 Ranking of three provinces in Northeast China in terms of GDP growth in 2014

Province	GDP growth (%)	Rankings
Liaoning	5.8	29
Jilin	6.5	27
Heilongjiang	5.6	30

Table A.12 Ranking of three provinces in Northeast China in terms of GDP growth in the first three quarters of 2015

Province	GDP growth (%)	Rankings
Liaoning	2.7	31
Jilin	6.3	28
Heilongjiang	5.5	29

Table A.13 Rankings of China's top 50 enterprises in 2014

Rankings	Company name	Operating revenue/million yuan	Profit/million yuan	Location of headquarters
1	China Petroleum and Chemical Corporation	2,880,311	67,179	Beijing
2	PetroChina Co., Ltd.	2,258,124	129,577	Beijing
3	China State Construction Engineering Corp.	681,048	20,399	Beijing
4	China Mobile Communications Corporation	630,177	121,692	Hong Kong
5	Industrial And Commercial Bank of China	589,637	262,649	Beijing

(continued)

Table A.13 (continued)

Rankings	Company name	Operating revenue/million yuan	Profit/million yuan	Location of headquarters
6	China Railway Construction Corporation Limited	586,790	10,345	Beijing
7	SAIC Motor Corporation Limited	565,807	24,804	Shanghai
8	China Railway Group Limited	560,444	9,375	Beijing
9	China Construction Bank	508,608	214,657	Beijing
10	Agricultural Bank of China	462,625	166,315	Beijing
11	China Life Insurance Company Limited	423,613	24,765	Beijing
12	Bank of China	407,508	156,911	Beijing
13	Ping An Insurance (Group) Company of China, Ltd.	362,631	28,154	Shenzhen
14	China Construction Communications Engrg. Group Ltd.	332,487	12,139	Beijing
15	China Telecom Corporation Limited	321,584	17,545	Beijing
16	The People's Insurance Company (Group)	304,738	8,121	Beijing
17	China United Network Communications Limited	303,727	3,443	Shanghai
18	CNOOC Limited	285,857	56,461	Hong Kong
19	China Shenhua Energy Company Limited	283,797	45,678	Beijing
20	Lenovo China	235,993	4,983	Hong Kong
21	The People's Insurance Company (Group) of China Limited	223,525	10,558	Beijing
22	Minmetals Development Co., Ltd.	203,259	280	Beijing

(continued)

Table A.13 (continued)

Rankings	Company name	Operating revenue/million yuan	Profit/million yuan	Location of headquarters
23	Metallurgical Corp. of China Ltd.	202,690	2,981	Beijing
24	China Pacific Insurance(group) Co., Ltd.	193,137	9,261	Shanghai
25	Baoshan Iron and Steel Co., Ltd.	190,026	5,818	Shanghai
26	Jiangxi Copper Company Limited	175,890	3,565	Guixi
27	Aluminum Corporation of China Limited	173,038	948	Beijing
28	Sinopharm Group Co., Ltd.	166,866	2,250	Shanghai
29	Bank of Communications	164,435	62,295	Shanghai
30	China Electric Power Construction Corp.	144,837	4,556	Beijing
31	China Vanke Co., Ltd.	135,419	15,119	Shenzhen
32	Huaneng Power Intl Inc.	133,833	10,520	Beijing
33	China Merchants Bank	132,604	51,743	Shenzhen
34	New China Life Insurance Company Ltd.	129,594	4,422	Beijing
35	Midea Group	121,265	5,317	Foshan
36	Gree Electric Appliances, Inc.	120,043	10,871	Zhuhai
37	China National Building Material Company Limited	117,688	5,762	Beijing
38	China Minsheng Bank	115,886	42,278	Beijing
39	Sinopec Shanghai Petrochemical Company Limited	115,540	2,004	Shanghai
40	China Resources Enterprise Limited	115,114	1,500	Hong Kong

(continued)

Table A.13 (continued)

Rankings	Company name	Operating revenue/million yuan	Profit/million yuan	Location of headquarters
41	Hebei Iron and Steel Co., Ltd.	110,255	116	Shijia zhuang
42	Industrial Bank Co., Ltd.	109,287	41,211	Fuzhou
43	Suning Commerce Group Co., Ltd.	105,292	372	Nanjing
44	Shanxi Taigang Stainless Steel Co., Ltd.	105,020	630	Taiyuan
45	China Citic Bank	104,558	39,175	Beijing
46	Xiamen C&D Inc.	102,068	2,693	Xiamen
47	Shanghai Construction Group Co., Ltd.	102,036	1,618	Shanghai
48	Shanghai Pudong Development Bank	100,015	40,922	Shanghai
49	China Southern Airlines	98,130	1,895	Guangzhou
50	China South Locomotive and Rolling Stock Corporation Limited	97,886	4,140	Beijing

Table A.14 China's top 50 publicly listed enterprises in 2014

Rankings	Name	Industry	Capital stock/100 million	Total market value/100 million yuan
1	PetroChina	Mining	1,830.21	13,799.78
2	ICBC	Bank	3,514.05	11,912.63
3	CCB	Bank	2,500.11	10,325.45
4	ABC	Bank	3,247.94	8,184.81
5	BOC	Bank	2,793.65	7,123.80
6	Sinopec	Chemical	1,167.96	6,155.13
7	China Life	Non-bank financial	282.65	3,846.83
8	Ping An Insurance	Non-bank financial	79.16	3,114.21
9	China Shenhua	Mining	198.90	2,891.95

(continued)

Table A.14 (continued)

Rankings	Name	Industry	Capital stock/100 million	Total market value/100 million yuan
10	BOCOM	Bank	742.63	2,881.39
11	China Merchants Bank	Bank	252.20	2,582.51
12	CMBC	Bank	340.40	2,113.86
13	China CITIC	Bank	467.87	1,997.82
14	Industrial Bank	Bank	190.52	1,910.95
15	Shanghai Pudong Development Bank	Bank	186.53	1,688.14
16	SAIC Motor	Automobile	110.26	1,686.91
17	Kweichow Moutai	Food and Beverage	11.42	1,621.41
18	CPIC	Non-bank financial	90.62	1,612.13
19	CITIC	Non-bank financial	110.17	1,262.54
20	China Everbright Bank	Bank	466.79	1,185.65
21	BYD	Automobile	24.76	1,175.85
22	Ping An Bank	Bank	114.25	1,132.21
23	China Yangtze Power	Public utilities	165.00	1,026.30
24	SIPG	Transportation	227.55	1,012.61
25	Daqin Railway	Transportation	148.67	938.09
26	China Vanke	Real estate	110.15	910.94
27	Gree Electric	Home appliances	30.08	885.82
28	Haitong Securities	Non-bank financial	95.85	877.00
29	CSCEC	Architectural ornament	300.00	846.00
30	China Shipbuilding	Defence industry	175.31	845.00
31	COSL	Mining	47.72	839.32
32	Anhui Conch Cement	Building materials	52.99	833.05
33	Midea	Home appliances	42.16	814.49
34	Huaneng Power International	Public utilities	140.55	795.53

(continued)

Table A.14 (continued)

Rankings	Name	Industry	Capital stock/100 million	Total market value/100 million yuan
35	Shuanghui Investment	Food and Beverage	22.01	787.59
36	Great Wall Motor	Automobile	30.42	769.73
37	BOE	Electronics	352.90	765.79
38	Huaxia Bank	Bank	89.05	730.18
39	Bank of Beijing	Bank	88.00	709.29
40	China Unicom	Communication	211.97	684.65
41	Wuliangye	Food and Beverage	37.96	680.62
42	Hikvision	Computer	40.17	680.52
43	Yili	Food and Beverage	20.43	676.61
44	Baoshan Iron and Steel	Steel	164.72	675.34
45	New China Life Insurance	Non-bank financial	31.20	659.16
46	CSR	machinery	138.03	621.14
47	China Communications Construction	Architectural ornament	161.75	608.17
48	China Merchants Securities	Non-bank financial	58.08	587.20
49	GF Securities	Non-bank financial	59.19	580.09
50	Changan Automobile	Automobile	46.63	574.00

Sourcing from statistics of stock market

The manufacturer's authorised representative in the EU is Springer Nature Customer Service Centre GmbH, Europaplatz 3, 69115 Heidelberg, Germany. If you have any concerns regarding our products, please contact ProductSafety@springernature.com

Printed and bound by CPI Group (UK) Ltd, Croydon, CR0 4YY
25/09/2026
02222903-0006